D0037273

Programming the
Raspberry Pi™

About the Author

Dr. Simon Monk (Preston, UK) has a degree in cybernetics and computer science and a Ph.D. in software engineering. Simon spent several years as an academic before he returned to the industry, co-founding the mobile software company Momote Ltd. Simon is now a full-time author and has published three books in the McGraw-Hill *Evil Genius* series. He is also the author of *Programming Arduino* and has published books on IOIO and .NET Gadgeteer. You can follow Simon on Twitter @simonmonk2.

Programming the Raspberry Pi™
Getting Started with Python

Simon Monk

New York   Chicago   San Francisco
Lisbon   London   Madrid   Mexico City
Milan   New Delhi   San Juan
Seoul   Singapore   Sydney   Toronto

Cataloging-in-Publication Data is on file with the Library of Congress

McGraw-Hill books are available at special quantity discounts to use as premiums and sales promotions, or for use in corporate training programs. To contact a representative, please e-mail us at bulksales@mcgraw-hill.com.

Programming the Raspberry Pi™: Getting Started with Python

4 5 6 7 8 9 0   DOC DOC   1 0 9 8 7 6 5 4 3

ISBN   978-0-07-180783-8

MHID   0-07-180783-7

Sponsoring Editor	**Indexer**
Roger Stewart	Jack Lewis
Editorial Supervisor	**Production Supervisor**
Patty Mon	George Anderson
Project Manager	**Composition**
Vastavikta Sharma, Cenveo Publisher Services	Cenveo Publisher Services
	Illustration
Copy Editor	Cenveo Publisher Services
Bart Reed	**Art Director, Cover**
Proofreader	Jeff Weeks
Carol Shields	

To my brothers, Andrew and Tim Monk, for their love and wisdom.

CONTENTS AT A GLANCE

CONTENTS

ACKNOWLEDGMENTS

As always, I thank Linda for her patience and support.

I also thank Andrew Robinson and my son, Matthew Monk, for their technical review of much of the material in this book. Check out Andrew's Raspberry Pi project book. I'm sure it will be excellent.

From TAB/McGraw-Hill, my thanks go out to my patient and thoroughly nice editor Roger Stewart and the excellent project management of Vastavikta Sharma and Patty Mon. It is always a pleasure to work with such a great team.

INTRODUCTION

The Raspberry Pi is rapidly becoming a worldwide phenomena. People are waking up to the possibility of a $35 (U.S.) computer that can be put to use in all sorts of settings—from a desktop workstation to a media center to a controller for a home automation system.

This book explains in simple terms, to both nonprogrammers and programmers new to the Raspberry Pi, how to start writing programs for the Pi in the popular Python programming language. It then goes on to give you the basics of creating graphical user interfaces and simple games using the pygame module.

The software in the book mostly uses Python 3, with the occasional use of Python 2 where necessary for module availability. The Raspbian Wheezy distribution recommended by the Raspberry Pi Foundation is used throughout the book.

The book starts with an introduction to the Raspberry Pi and covers the topics of buying the necessary accessories and setting everything up. You then get an introduction to programming while you gradually work your way through the next few chapters. Concepts are illustrated with sample applications that will get you started programming your Raspberry Pi.

Three chapters are devoted to programming and using the Raspberry Pi's GPIO connector, which allows the device to be attached to external electronics. These chapters include two sample projects—one for making an LED clock and the other a Raspberry Pi controller robot, complete with ultrasonic rangefinder.

Here are the key topics covered in the book:

- Python numbers, variables, and other basic concepts
- Strings, lists, dictionaries, and other Python data structures
- Modules and object orientation
- Files and the Internet
- Graphical user interfaces using Tkinter

- Game programming using Pygame

- Interfacing with hardware via the GPIO connector

- Sample hardware projects

All the code listings in the book are available for download from the book's website at http://www.raspberrypibook.com, where you can also find other useful material relating to the book, including errata.

1

Introduction

The Raspberry Pi went on general sale at the end of February 2012 and immediately crashed the websites of the suppliers chosen to take orders for it. So what was so special about this little device and why has it created so much interest?

What Is the Raspberry Pi?

The Raspberry Pi, shown in Figure 1-1, is a computer that runs the Linux operating system. It has USB sockets you can plug a keyboard and mouse into and HDMI (High-Definition Multimedia Interface) video output you can connect a TV or monitor into. Many monitors only have a VGA connector, and Raspberry Pi will not work with this. However, if your monitor has a DVI connector, cheap HDMI-to-DVI adapters are available.

When Raspberry Pi boots up, you get the Linux desktop shown in Figure 1-2. This really is a proper computer, complete with an office suite, video playback capabilities, games, and the lot. It's not Microsoft Windows; instead, it is Windows open source rival Linux (Debian Linux), and the windowing environment is called LXDE.

Its small (the size of a credit card) and extremely affordable (starting at $25). Part of the reason for this low cost is that some components are not included with the board or are optional extras. For instance, it does not come in a case to protect it—it is just a bare board. Nor does it come with a power supply, so you will need to find yourself a 5V micro-USB power supply, much like you would use to charge a phone (but probably with higher power). A USB power supply and a micro-USB lead are often used for this.

Figure 1-1   *The Raspberry Pi*

Figure 1-2   *The Raspberry Pi desktop*

What Can You Do with a Raspberry Pi?

You can do pretty much anything on a Raspberry Pi that you can on any other Linux desktop computer, with a few limitations. The Raspberry Pi uses an SD card in place of a hard disk, although you can plug in a USB hard disk. You can edit office documents, browse the Internet, and play games (even games with quite intensive graphics, such as *Quake*).

The low price of the Raspberry Pi means that it is also a prime candidate for use as a media center. It can play video, and you can just about power it from the USB port you find on many TVs.

A Tour of the Raspberry Pi

Figure 1-3 labels the various parts of a Raspberry Pi. This figure takes you on a tour of the Model B Raspberry Pi, which differs from the Model A by virtue of having an RJ-45 LAN connector, allowing it to be connected to a network.

The RJ-45 Ethernet connector is shown in the top-left corner of the figure. If your home hub is handy, you can plug your Raspberry Pi directly into your

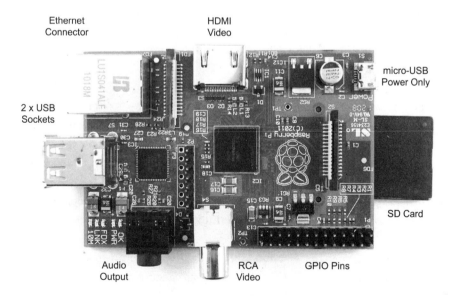

Figure 1-3 *The anatomy of a Raspberry Pi*

local network. While we are on the subject, it is worth noting that the Raspberry Pi does not have Wi-Fi built in. For wireless networking, you will need to plug in a USB wireless adapter. This may then require some additional work installing drivers.

Immediately below the Ethernet socket you'll find a pair of USB sockets, one on top of the other. You can plug a keyboard, mouse, or external hard disks into the board, but you'll fairly rapidly run out of sockets. For this reason, many people use a USB hub to gain a few more USB sockets.

In the bottom-left corner of the figure you'll find an audio socket that provides a stereo analog signal for headphones or powered speakers. The HDMI connector is also sound capable.

Next to the audio socket is an RCA video connector. You are unlikely to use this connector unless you are using your Raspberry Pi with an older TV. You are far more likely to use the HDMI connector immediately opposite it, shown at the top of the figure. HDMI is higher quality, includes sound, and can be connected to DVI-equipped monitors with a cheap adapter.

To the right of the yellow RCA jack are two rows of pins. These are called GPIO (General Purpose Input/Output) pins, and they allow the Raspberry Pi to be connected to custom electronics. Users of the Arduino and other microcontroller boards will be used to the idea of GPIO pins. Later, in Chapter 11, we will use these pins to enable our Raspberry Pi to be the "brain" of a little roving robot by controlling its motors. In Chapter 10, we will use the Raspberry Pi to make an LED clock.

The Raspberry Pi has an SD card slot underneath the board. This SD card needs to be at least 2GB in size. It contains the computer's operating system as well as the file system in which you can store any documents you create. The SD card is an optional extra feature when buying your Raspberry Pi. Preparing your own SD card is a little complex to do, and suppliers such as SK Pang, Farnell, and RS Components all sell already-prepared SD cards. Because no disk is built into your Raspberry Pi, this card is effectively your computer, so you could take it out and put it in a different Raspberry Pi and all your stuff would be there.

Above the SD card is a micro-USB socket. This is only used to supply power to the Raspberry Pi. Therefore, you will need a power supply with a micro-USB connector on the end. This is the same type of connector used by many mobile phones, including most Android phones. Do, however, check that it is capable of supplying at least 700mA; otherwise, your Raspberry Pi may behave erratically.

For those interested in technical specs, the big square chip in the center of the board is where all the action occurs. This is Broadcom's "System on a Chip" and includes 256MB of memory as well as the graphics and general-purpose processors that drive the Raspberry Pi.

You may also have noticed flat cable connectors next to the SD card and between the Ethernet and HDMI connectors. These are for LCD displays and a camera, respectively. Look for camera and LCD display modules becoming available for the Pi in the near future.

Setting Up Your Raspberry Pi

You can make your life easier by buying a prepared SD card and power supply when you buy your Raspberry Pi, and for that matter you may as well get a USB keyboard and mouse (unless you have them lurking around the house somewhere). Let's start the setup process by looking at what you will need and where to get it from.

Buying What You Need

Table 1-1 shows you what you will need for a fully functioning Raspberry Pi system. At the time of writing, the Raspberry Pi itself is sold through two worldwide distributors based in the UK: Farnell (and the related U.S. company Newark) and RS Components, which is not to be confused with RadioShack.

Power Supply

Figure 1-4 show a typical USB power supply and USB-A-to-micro-USB lead.

You may be able to use a power supply from an old MP3 player or the like, as long as it is 5V and can supply enough current. It is important not to overload the power supply because it could get hot and fail (or even be a fire hazard). Therefore, the power supply should be able to supply at least 700mA, but 1A would give the Raspberry Pi a little extra when it comes to powering the devices attached to its USB ports.

If you look closely at the specs written on the power supply, you should be able to determine its current supply capabilities. Sometimes its power-handling

Item	Source and Part Number	Additional Information
Raspberry Pi, Model A or B	Farnell (www.farnell.com) Newark (www.newark.com) RS Components (www .rs-components.com)	The difference between the two models is that the Model B has a network connection.
USB power supply (U.S. mains plug)	Newark: 39T2392 RadioShack: 55053163 Adafruit PID:501	5V USB power supply. Should be capable of supplying 700mA (3W), but 1A (5W) is better.
USB power supply (UK mains plug)	Farnell: 2100374 Maplins: N15GN	
USB power supply (European mains plug)	Farnell: 1734526	
Micro-USB lead	RadioShack: 55048949 Farnell: 2115733 Adafruit PID 592	
Keyboard and mouse	Any computer store	Any USB keyboard will do. Also, wireless keyboards and mice that come with their own USB adaptor will work, too.
TV/monitor with HDMI	Any computer/electrical store	
HDMI lead	Any computer/electrical store	
SD card (prepared)	SK Pang: RSP-2GBSD Newark: 96T7436 Farnell: 2113756	
Wi-Fi adapter*	http://elinux.org/RPi_ VerifiedPeripherals#USB_WiFi_ Adapters	Elinux.org provides an up-to-date list of Wi-Fi adapters.
USB hub*	Any computer store	
HDMI-to-DVI adapter*	Newark: 74M6204 Maplins: N24CJ Farnell: 1428271	
Ethernet patch cable*	Any computer store	
Case*	Adafruit, SK Pang, or Alliedelec .com	
*These items are optional.		

Table 1-1 *A Raspberry Pi Kit*

capabilities will be expressed in watts (W); if that's the case, it should be at least 3W. If it indicates 5W, this is equivalent to 1A.

Keyboard and Mouse

The Raspberry Pi will work with pretty much any USB keyboard and mouse. You can also use most wireless USB keyboards and mice—the kind that come

Figure 1-4 *USB power supply*

with their own dongle to plug into the USB port. This is quite a good idea, especially if they come as a pair. That way, you are only using up one of the USB ports. This will also come in quite handy in Chapter 10 when we use a wireless keyboard to control our Raspberry Pi–based robot.

Display

Including an RCA video output on the Raspberry Pi is, frankly, a bit puzzling because most people are going to go straight to the more modern HDMI connector. A low-cost 22-inch LCD TV will make a perfectly adequate display for the Pi. Indeed, you may just decide to use the main family TV, just plugging the Pi into the TV when you need it.

If you have a computer monitor with just a VGA connector, you are not going to be able to use it without an expensive converter box. On the other hand, if your monitor has a DVI connector, an inexpensive adapter will do the job well.

SD Card

You can use your own SD card in the Raspberry Pi, but it will need to be prepared with an operating system disk image. This is a little fiddly, so you may just want to spend a dollar or two more and buy an SD card that is already prepared and ready to go.

You can also find people at Raspberry Pi meet-ups who will be happy to help you prepare an SD card. The prepared SD cards supplied by Farnell and RS Components are overpriced. Look around on the Internet to find suppliers (such as SK Pang) who sell prepared cards, with the latest operating system distribution, for less than you would pay for an SD card in a supermarket. If you indeed want to "roll your own" SD card, refer to the instructions found at www.raspberrypi.org/downloads.

To prepare your own card, you must have another computer with an SD card reader. The procedure is different depending on whether your host computer is a Windows, Mac, or Linux machine. However, various people have produced useful tools that try to automate the process as much as possible.

If you decide to roll your own, be sure to follow the instructions carefully—with some tools, it is quite easy to accidentally reformat a hard disk attached to your computer if the tool mistakes it for the SD card! Fortunately, this process is getting better all the time as easier-to-use software tools become available.

A big advantage of making your own SD card is that you can actually choose from a range of operating system distributions. Table 1-2 shows the most popular ones available at the time of writing. Check on the Raspberry Pi Foundation's website for newer distributions.

Of course, nothing is stopping you from buying a few SD cards and trying out the different distributions to see which you prefer. However, if you are a Linux beginner, you should stick to the standard Raspbian Wheezy distribution.

Case

The Raspberry Pi does not come in any kind of enclosure. This helps to keep the price down, but also makes it rather vulnerable to breakage. Therefore, it is a good idea to either make or buy a case as soon as you can. Figure 1-5 shows a few of the ready-made cases currently available.

The cases shown are supplied by (a) Adafruit (www.adafruit.com), (b) SK Pang (www.skpang.co.uk/), and (c) ModMyPi (www.modmypi.com). The case

Distribution	Notes
Raspbian Wheezy	This is the "standard" Raspberry Pi operating system and the one used in all the examples in this book. It uses the LXDE desktop.
Arch Linux ARM	This distribution is more suited to Linux experts.
QtonPi	This distribution is intended for people developing rich graphical programs using the Qt5 graphics framework.
Occidentalis	A distribution made by Adafruit and based on Raspbian Wheezy but with improvements intended for hardware hackers.

Table 1-2 *Raspberry Pi Linux Distributions*

(a)   (b)   (c)

Figure 1-5 *Commercial Raspberry Pi cases*

you choose will depend on what you plan to do with your Raspberry Pi. If you have access to a 3D printer, you can also use the following open source designs:

- www.thingiverse.com/thing:23446
- www.thingiverse.com/thing:24721

You can also find a folded card design called the Raspberry Punnet at www .raspberrypi.org/archives/1310.

People are having a lot of fun building their Raspberry Pi into all sorts of repurposed containers, such as vintage computers and games consoles. One could even build a case using Legos. My first case for a Raspberry Pi was made by cutting holes in a plastic container that used to hold business cards (see Figure 1-6).

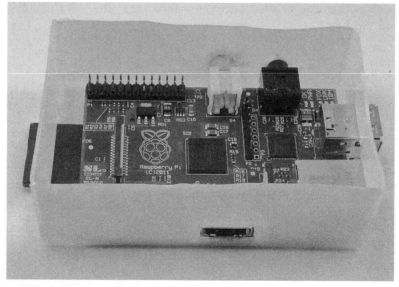

Figure 1-6 *A homemade Raspberry Pi case*

Wi-Fi

Neither of the Raspberry Pi models has support for Wi-Fi. Therefore, to wirelessly connect your Raspberry Pi to the network, you have just two options. The first is to use a USB wireless adapter that just plugs into a USB socket (see Figure 1-7). With any luck, Linux should recognize it and immediately allow you to connect (or show what you need to do to connect).

The Wi-Fi adapters in the list referenced in Table 1-1 are purported to work with the Raspberry Pi. However, there are sometimes problems with Wi-Fi drivers, so be sure to check the Raspberry Pi forum and wiki for up-to-date information on compatible devices.

The second option for Wi-Fi is to use a Wi-Fi bridge with a Model B Raspberry Pi. These devices are usually USB powered and plug into the Ethernet socket on the Raspberry Pi. They are often used by the owners of game consoles that have an Ethernet socket but no Wi-Fi. This setup has the advantage in that the Raspberry Pi does not require any special configuration.

USB Hub

Because the Raspberry Pi has just two USB ports available, you will rapidly run out of sockets. The way to obtain more USB ports is to use a USB hub (see Figure 1-8).

These hubs are available with anywhere from three to eight ports. Make sure that the port supports USB 2. It is also a good idea to use a "powered" USB hub so that you do not draw too much power from the Raspberry Pi.

Figure 1-7 *Wi-Fi adapter*

Figure 1-8   *A USB hub*

Connecting Everything Together

Now that you have all the parts you need, let's get it all plugged together and boot your Raspberry Pi for the first time. Figure 1-9 shows how everything needs to be connected.

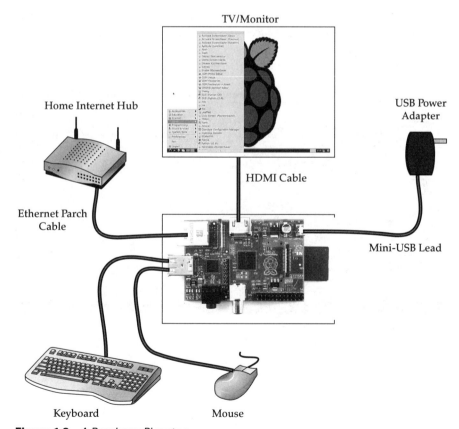

Figure 1-9   *A Raspberry Pi system*

Insert the SD card, connect the keyboard, mouse, and monitor to the Pi, attach the power supply, and you are ready to go.

Booting Up

The first time you boot your Raspberry Pi, it will not immediately boot into the kind of graphical environment you would normally see in, say, a Windows computer. Instead, it will stop to allow a first-time configuration (see Figure 1-10). It is a good idea to make a number of the configuration changes shown here.

First, if your SD card is larger than 2GB, the Raspberry Pi will only make use of the first 2GB unless you select the option to expand_rootfs. Select this option using the UP and DOWN ARROW keys and ENTER.

Another change well worth making is the boot_behaviour option. If this is not set to Boot Straight to Desktop, you will be forced to log in and start the windowing environment manually each time you power up your Raspberry Pi (see Figure 1-11).

Figure 1-10 *Configuration screen*

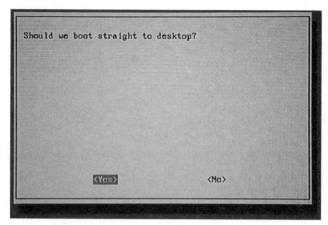

Figure 1-11   *Boot-to-desktop option*

Summary

Now that we have set up our Raspberry Pi and it is ready to use, we can start exploring some of its features and get a grip on the basics of Linux.

2

Getting Started

The Raspberry Pi uses Linux as its operating system. This chapter introduces Linux and shows you how to use the desktop and command line.

Linux

Linux is an open source operating system. This software has been written as
a community project for those looking for an alternative to the duopoly of
Microsoft Windows and Apple OS X. It is a fully featured operating system
based on the same solid UNIX concepts that arose in the early days of com-
puting. It has a loyal and helpful following and has matured into an operating system that is powerful and easy to use.

Although the operating system is called Linux, various Linux distributions (or *distros)* have been produced. These involve the same basic operating
system, but are packaged with different bundles of applications or different
windowing systems. Although many distros are available, the one recommended by the Raspberry Pi foundation is called Raspbian Wheezy.

If you are only used to some flavor of Microsoft Windows, expect to expe-
rience some frustration as you get used to a new operating system. Things
work a little differently in Linux. Almost anything you may want to change
about Linux can be changed. The system is open and completely under your
control. However, as they say in *Spiderman,* with great power comes great
responsibility. This means that if you are not careful, you could end up break-
ing your operating system.

The Desktop

At the end of Chapter 1, we had just booted up our Raspberry Pi, logged in, and started up the windowing system. Figure 2-1 serves to remind you of what the Raspberry Pi desktop looks like.

If you are a user of Windows or Mac computers, you will be familiar with the concept of a desktop as a folder within the file system that acts as a sort of background to everything you do on the computer.

Along the left side of the desktop, you see some icons that launch applications. Clicking the left-most icon on the bar at the bottom of the screen will show us some of the applications and tools installed on the Raspberry Pi (rather like the Start menu in Microsoft Windows). We are going to start with the File Manager, which can be found under the Accessories.

The File Manager is just like the File Explorer in Windows or the Finder on a Mac. It allows you to explore the file system, copy and move files, as well as launch files that are executable (applications).

Figure 2-1   *Raspberry Pi desktop*

When it starts, the File Manager shows you the contents of your home directory. You may remember that when you logged in, you gave your login name as pi. The root to your home directory will be /home/pi. Note that like Mac's OS X, Linux uses slash (/) characters to separate the parts of a directory name. Therefore, / is called the *root* directory and /home/ is a directory that contains other directories, one for each user. Our Raspberry Pi is just going to have one user (called pi), so this directory will only ever contain a directory called pi. The current directory is shown in the address bar at the top, and you can type directly into it to change the directory being viewed, or you can use the navigation bar at the side. The contents of the directory /home/pi include just the directories Desktop and python_games.

Double-clicking Desktop will open the Desktop directory, but this is not of much interest because it just contains the shortcuts on the left side of the desktop. If you open python_games, you will see some games you can try out, as shown in Figure 2-2.

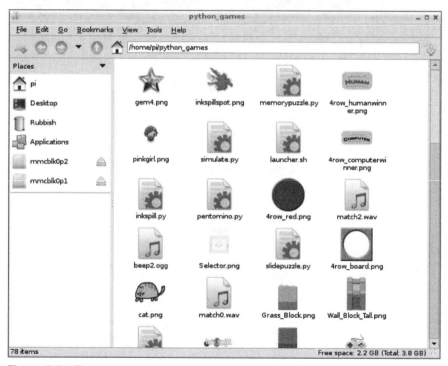

Figure 2-2   *The contents of python_games, as shown in File Manager*

You shouldn't often need to use any of the file system outside of your home directory. You should keep all documents, music files, and so on, housed within directories on your home folder or on an external USB flash drive.

The Internet

If you have a home hub and can normally plug in any Internet device using an Ethernet cable, you should have no problem getting your Raspberry Pi online. Your home hub should automatically assign the Raspberry Pi an IP address and allow it to connect to the network.

The Raspberry Pi comes with a web browser called Midori, which you will find under the Internet section of your start menu. You can check that your connection is okay by starting Midori and connecting to a website of your choice, as shown in Figure 2-3.

Figure 2-3 *The Midori web browser*

The Command Line

If you are a Windows or Mac user, you may have never used the command line. If you are a Linux user, on the other hand, you almost certainly will have done so. In fact, if you are a Linux user, then about now you will have realized that you probably don't need this chapter because it's all a bit basic for you.

Although it is possible to use a Linux system completely via the graphical interface, in general you will need to type commands into the command line. You do this to install new applications and to configure the Raspberry Pi.

From the launcher button (bottom left), open the LXTerminal, which is shown in Figure 2-4.

Navigating with the Terminal

You will find yourself using three commands a lot when you are using the command line. The first command is pwd, which is short for *print working*

Figure 2-4 *The LXTerminal command line*

directory and shows you which directory you are currently in. Therefore, after the $ sign in the terminal window, type **pwd** and press RETURN, as shown in Figure 2-5.

As you can see, we are currently in /home/pi. Rather than provide a screen shot for everything we are going to type into the terminal, I will use the convention that anything I want you to type will be prefixed with a $ sign, like this:

```
$pwd
```

Anything you should see as a response will not have $ in front of it. Therefore, the whole process of running the pwd command would look something like this:

```
$pwd
/home/pi
```

The next common command we are going to discuss is ls, which is short for *list* and shows us a list of the files and directories within the working directory. Try the following:

```
$ls
Desktop
```

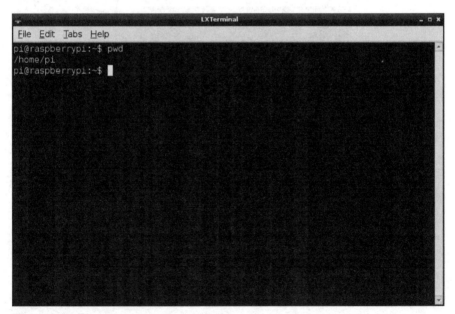

Figure 2-5   *The pwd command*

This tells us that the only thing in /home/pi is the directory Desktop.

The final command we are going to cover for navigating around is cd (which stands for *change directory*). This command changes the current working directory. It can change the directory relative either to the old working directory or to a completely different directory if you specify the whole directory, starting with /. So, for example, the following command will change the current working directory to /home/pi/Desktop:

```
$pwd
/home/pi
$cd Desktop
```

You could do the same thing by typing this:

```
cd /home/pi/Desktop
```

Note that when entering a directory or filename, you do not have to type all of it. Instead, at any time after you have typed some of the name, you can press the TAB key. If the filename is unique at that point, it will be automatically completed for you.

sudo

Another command that you will probably use a lot is sudo (for super-user do). This runs whatever command you type after it as if you were a super-user. You might be wondering why, as the sole user of this computer, you are not automatically a super-user. The answer is that, by default, your regular user account (username: pi, password: raspberry) does not have privileges that, say, allow you to go to some vital part of the operating system and start deleting files. Instead, to cause such mayhem, you have to prefix those commands with sudo. This just adds a bit of protection against accidents.

For the commands we have discussed so far, you will not need to prefix them with sudo. However, just for interest, try typing the following:

```
sudo ls
```

This will work the same way ls on its own works; you are still in the same working directory. The only difference is that you will be asked for your password the first time you use sudo.

Applications

The Raspbian Wheezy distribution for Raspberry Pi is fairly sparse. However, loads of applications can be installed. Installing new applications requires the command line again. The command `apt-get` is used to both install and uninstall applications. Because installing an application often requires superuser privileges, you should prefix `apt-get` commands with `sudo`.

The command `apt-get` uses a database of available packages that is updated over the Internet, so the first `apt-get` command you should use is

```
sudo apt-get update
```

which updates the database of packages. You will need to be connected to the Internet for it to work.

To install a particular package, all you need to know is the package manager name for it. For example, to install the Abiword word processor application, all you need to type is the following:

```
sudo apt-get install abiword
```

It will take a while for everything that is needed to be downloaded and installed, but at the end of the process you will find that you have a new folder in your start menu called Office that contains the application Abiword (see Figure 2-6).

You will notice that the text document in Abiword is actually part of this chapter. In fact, it is close to this part of this chapter, as I am writing it. (I can feel myself falling into a recursive hole. I may well vanish in a puff of logic.)

Abiword is a perfectly serviceable word processor. If I didn't love my Mac quite so much, I would be tempted to write this entire book on my Raspberry Pi.

While we are on the subject of office applications, the spreadsheet stable mate of Abiword is called Gnumeric. To install it, here is all you need to type:

```
sudo apt-get install gnumeric
```

Once this application is installed, another option will have appeared in your Office menu—this one for Gnumeric.

To find out about other packages you might want to install, look for recommendations on the Internet, especially on the Raspberry Pi forum (www.raspberrypi.org/phpBB3). You can also browse the list of packages available for Raspbian Wheezy at http://packages.debian.org/stable/.

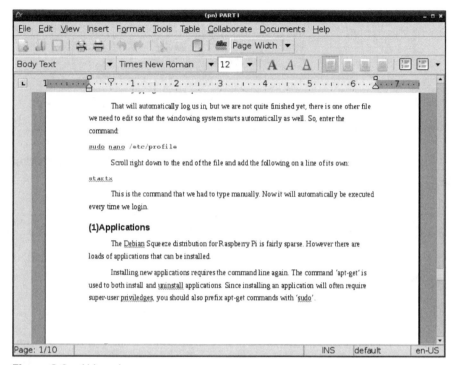

Figure 2-6 *Abiword screen*

Not all of these packages will work, because the Raspberry Pi does not have vast amounts of memory and storage available to it; however, many will.

If you want to remove a package, use the following command:

```
sudo apt-get remove --auto-remove --purge packagename
```

This removes both the package and any packages it depends on that are not used by something else that still needs them. Be sure to keep an eye on the bottom-right corner of your File Manager window; it will tell you how much free space is available.

Internet Resources

Aside from the business of programming the Raspberry Pi, you now have a functioning computer that you are probably keen to explore. To help you with this, many useful Internet sites are available where you can obtain advice and recommendations for getting the most out of your Raspberry Pi.

Site	Description
www.raspberrypi.org	The home page of the Raspberry Pi Foundation. Check out the forum and FAQs.
www.raspberrypi-spy.co.uk	A blog site with useful how-to posts.
http://elinux.org/RaspberryPiBoard	The main Raspberry Pi wiki. Lots of information about the Raspberry Pi, especially a useful list of verified peripherals (http://elinux.org/RPi_VerifiedPeripherals).

Table 2-1   *Internet Resources for the Raspberry Pi*

Table 2-1 lists some of the more useful sites relating to the Raspberry Pi. Your search engine will happily show you many more.

Summary

Now that we have everything set up and ready to go on our Raspberry Pi, it is time to start programming in Python.

3

Python Basics

The time has come to start creating some of our own programs for the Raspberry Pi. The language we are going to use is called Python. It has the great benefit that it is easy to learn while at the same time being powerful enough to create some interesting programs, including some simple games and programs that use graphics.

As with most things in life, it is necessary to learn to walk before you can run, and so we will begin with the basics of the Python language.

Okay, so a programming language is a language for writing computer programs in. But why do we have to use a special language anyway? Why couldn't we just use a human language? How does the computer use the things that we write in this language?

The reason why we don't use English or some other human language is that human languages are vague and ambiguous. Computer languages use English words and symbols, but in a very structured way.

IDLE

The best way to learn a new language is to begin using it right away. So let's start up the program we are going to use to help us write Python. This program is called IDLE, and you will find it in the programming section of your start menu. In fact, you will find more than one entry for IDLE. Select the one labelled "IDLE 3" after it. Figure 3-1 shows IDLE and the Python Shell.

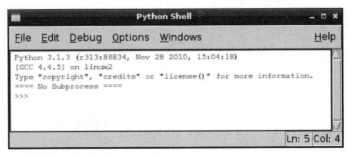

Figure 3-1   *IDLE and the Python Shell*

Python Versions

Python 3 was a major change over Python 2. This book is based on Python 3.1, but as you get further into Python you may find that some of the modules you want to use are not available for Python 3.

Python Shell

What you see in Figure 3-1 is the Python Shell. This is the window where you type Python commands and see what they do. It is very useful for little experiments, especially while you're learning Python.

Rather like at the command prompt, you can type in commands after the prompt (in this case, >>>) and the Python console will show you what it has done on the line below.

Arithmetic is something that comes naturally to all programming languages, and Python is no exception. Therefore, type **2 + 2** after the prompt in the Python Shell and you should see the result (4) on the line below, as shown in Figure 3-2.

```
                        Python Shell              _ □ x

 File   Edit   Debug   Options   Windows                Help

 Python 3.1.3 (r313:86834, Nov 28 2010, 15:04:18)
 [GCC 4.4.5] on linux2
 Type "copyright", "credits" or "license()" for more information.
 ==== No Subprocess ====
 >>> 2 + 2
 4
 >>>

                                               Ln: 7 Col: 4
```

Figure 3-2   *Arithmetic in the Python Shell*

Editor

The Python Shell is a great place to experiment, but it is not the right place to write a program. Python programs are kept in files so that you do not have to retype them. A file may contain a long list of programming language commands, and when you want to run all the commands, what you actually do is run the file.

The menu bar at the top of IDLE allows us to create a new file. Therefore, select File and then New Window from the menu bar. Figure 3-3 shows the IDLE Editor in a new window.

Type the following two lines of code into IDLE:

```
print('Hello')
print('World')
```

You will notice that the editor does not have the >>> prompt. This is because what we write here will not be executed immediately; instead, it will just be stored in a file until we decide to run it. If you wanted, you could use nano or some other text editor to write the file, but the IDLE editor integrates nicely with Python. It also has some knowledge of the Python language and can thus serve as a memory aid when you are typing out programs.

We need a good place to keep all the Python programs we will be writing, so open the File Browser from the start menu (its under Accessories). Right-click

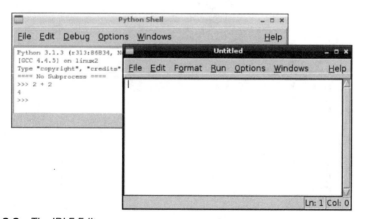

Figure 3-3 *The IDLE Editor*

Figure 3-4 *Creating a Python folder*

over the main area and select New and then Folder from the pop-up menu (see Figure 3-4). Enter the name **Python** for the folder and press the RETURN key.

Next, we need to switch back to our editor window and save the file using the File menu. Navigate to inside the new Python directory and give the file the name **hello.py**, as shown in Figure 3-5.

To actually run the program and see what it does, go to the Run menu and select Run Module. You should see the results of the program's execution in

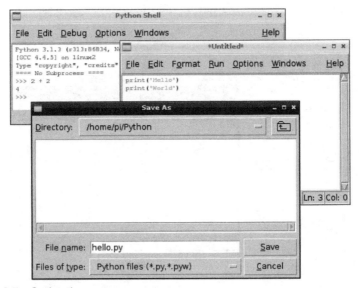

Figure 3-5 *Saving the program*

the Python Shell. It is no great surprise that the program prints the two words *Hello* and *World,* each on its own line.

What you type in the Python Shell does not get saved anywhere; therefore, if you exit IDLE and then start it up again, anything you typed in the Python Shell will be lost. However, because we saved our Editor file, we can load it at any time from the File menu.

NOTE *To save this book from becoming a series of screen dumps, from now on if I want you to type something in the Python Shell, I will proceed it with* >>>. *The results will then appear on the lines below it.*

Numbers

Numbers are fundamental to programming, and arithmetic is one of the things computers are very good at. We will begin by experimenting with numbers, and the best place to experiment is the Python Shell.

Type the following into the Python Shell:

```
>>> 20 * 9 / 5 + 32
68.0
```

This isn't really advancing much beyond the 2 + 2 example we tried before. However, this example does tell us a few things:

- * means multiply.

- / means divide.

- Python does multiplication before division, and it does division before addition.

If you wanted to, you could add some parentheses to guarantee that everything happens in the right order, like this:

```
>>> (20 * 9 / 5) + 32
68.0
```

The numbers you have there are all whole numbers (or *integers* as they are called by programmers). We can also use a decimal point if we want to use such numbers. In programming, these kinds of numbers are called *floats,* which is short for *floating point.*

Variables

Sticking with the numbers theme for a moment, let's investigate variables. You can think of a variable as something that has a value. It is a bit like using letters as stand-ins for numbers in algebra. To begin, try entering the following:

```
>>> k = 9.0 / 5.0
```

The equals sign assigns a value to a variable. The variable must be on the left side and must be a single word (no spaces); however, it can be as long as you like and can contain numbers and the underscore character (_). Also, characters can be upper- and lowercase. Those are the rules for naming variables; however, there are also conventions. The difference is that if you break the rules, Python will complain, whereas if you break the conventions, other programmers may snort derisively and raise their eyebrows.

The conventions for variables are that they should start with a lowercase letter and should use an underscore between what in English would be words (for instance, `number_of_chickens`). The examples in Table 3-1 give you some idea of what is legal and what is conventional.

Many other languages use a different convention for variable names called bumpy-case or camel-case, where the words are separated by making the start of each word (except the first one) uppercase (for example, `numberOfChickens`). You will sometimes see this in Python example code. Ultimately, if the code is just for your own use, then how the variable is written does not really matter, but if your code is going to be read by others, it's a good idea to stick to the conventions.

Variable Name	Legal	Conventional
x	Yes	Yes
X	Yes	No
number_of_chickens	Yes	Yes
number of chickens	No	No
numberOfChickens	Yes	No
NumberOfChickens	Yes	No
2beOrNot2b	No	No
toBeOrNot2b	Yes	No

Table 3-1 *Naming Variables*

By sticking to the naming conventions, it's easy for other Python programmers to understand your program.

If you do something Python doesn't like or understand, you will get an error message. Try entering the following:

```
>>> 2beOrNot2b = 1
SyntaxError: invalid syntax
```

This is an error because you are trying to define a variable that starts with a digit, which is not allowed.

A little while ago, we assigned a value to the variable k. We can see what value it has by just entering **k**, like so:

```
>>> k
1.8
```

Python has remembered the value of k, so we can now use it in other expressions. Going back to our original expression, we could enter the following:

```
>>> 20 * k + 32
68.0
```

For Loops

Arithmetic is all very well, but it does not make for a very exciting program. Therefore, in this section you will learn about *looping,* which means telling Python to perform a task a number of times rather than just once. In the following example, you will need to enter more than one line of Python. When you press RETURN and go to the second line, you will notice that Python is waiting. It has not immediately run what you have typed because it knows that you have not finished yet. The : character at the end of the line means that there is more to do.

These extra tasks must each appear on an indented line. Therefore, in the following program, at the start of the second line you'll press TAB once and then type **print(x)**. To get this two-line program to actually run, press RETURN twice after the second line is entered.

```
>>> for x in range(1, 10):
        print(x)
```

```
1
2
3
4
5
6
7
8
9
>>>
```

This program has printed out the numbers between 1 and 9 rather than 1 and 10. The `range` command has an exclusive end point—that it, it doesn't include the last number in the range, but it does include the first.

You can check this out by just taking the range bit of the program and asking it to show its values as a list, like this:

```
>>> list(range(1, 10))
[1, 2, 3, 4, 5, 6, 7, 8, 9]
```

Some of the punctuation here needs a little explaining. The parentheses are used to contain what are called *parameters*. In this case, `range` has two parameters: `from` (1) and `to` (10), separated by a comma.

The `for in` command has two parts. After the word `for` there must be a variable name. This variable will be assigned a new value each time around the loop. Therefore, the first time it will be 1, the next time 2, and so on. After the word `in`, Python expects to see something that works out to be a list of items. In this case, this is a list of the numbers between 1 and 9.

The `print` command also takes an argument that displays it in the Python Shell. Each time around the loop, the next value of x will be printed out.

Simulating Dice

We'll now build on what you just learned about loops to write a program that simulates throwing a die 10 times.

To do this, you will need to know how to generate a random number. So, first let's work out how to do that. If you didn't have this book, one way to find out how to generate a random number would be to type **random numbers python** into your search engine and look for fragments of code to type into the Python Shell. However, you do have this book, so here is what you need to write:

```
>>> import random
>>> random.randint(1,6)
2
```

Try entering the second line a few times, and you will see that you are getting different random numbers between 1 and 6.

The first line imports a library that tells Python how to generate numbers. You will learn much more about libraries later in this book, but for now you just need to know that we have to issue this command before we can start using the `randint` command that actually gives us a random number.

NOTE *I am being quite liberal with the use of the word* command *here. Strictly speaking, items such as* `randint` *are actually functions, not commands, but we will come to this later.*

Now that you can make a single random number, you need to combine this with your knowledge of loops to print off 10 random numbers at a time. This is getting beyond what can sensibly be typed into the Python Shell, so we will use the IDLE Editor.

You can either type in the examples from the text here or download all the Python examples used in the book from the book's website (www.raspberrypibook.com). Each programming example has a number. Thus, this program will be contained in the file 3_1_dice.py, which can be loaded into the IDLE Editor.

At this stage, it is worth typing in the examples to help the concepts sink in. Open up a new IDLE Editor window, type the following into it, and then save your work:

```
#3_1_dice
import random
for x in range(1, 11):
        random_number = random.randint(1, 6)
        print(random_number)
```

The first line begins with a # character. This indicates that the entire line is not program code at all, but just a comment to anyone looking at the program. Comments like this provide a useful way of adding extra information about a

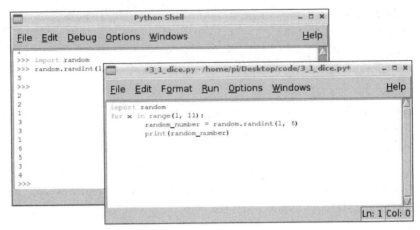

Figure 3-6   *The dice simulation*

program into the program file, without interfering with the operation of the program. In other words, Python will ignore any line that starts with #.

Now, from the Run menu, select Run Module. The result should look something like Figure 3-6, where you can see the output in the Python Shell behind the Editor window.

If

Now it's time to spice up the dice program so that two dice are thrown, and if we get a total of 7 or 11, or any double, we will print a message after the throw. Type or load the following program into the IDLE Editor:

```
#3_2_double_dice
import random
for x in range(1, 11):
        throw_1 = random.randint(1, 6)
        throw_2 = random.randint(1, 6)
        total = throw_1 + throw_2
        print(total)
        if total == 7:
                print('Seven Thrown!')
        if total == 11:
                print('Eleven Thrown!')
        if throw_1 == throw_2:
                print('Double Thrown!')
```

When you run this program, you should see something like this:

```
6
7
Seven Thrown!
9
8
Double Thrown!
4
4
8
10
Double Thrown!
8
8
Double Thrown!
```

The first thing to notice about this program is that now two random numbers between 1 and 6 are generated. One for each of the dice. A new variable, total, is assigned to the sum of the two throws.

Next comes the interesting bit: the if command. The if command is immediately followed by a condition (in the first case, total == 7). There is then a colon (:), and the subsequent lines will only be executed by Python if the condition is true. At first sight, you might think there is a mistake in the condition because it uses == rather than =. The double equal sign is used when comparing items to see whether they are equal, whereas the single equal sign is used when assigning a value to a variable.

The second if is not tabbed in, so it will be executed regardless of whether the first if is true. This second if is just like the first, except that we are looking for a total of 11. The final if is a little different because it compares two variables (throw_1 and throw_2) to see if they are the same, indicating that a double has been thrown.

Now, the next time you go to play *Monopoly* and find that the dice are missing, you know what to do: Just boot up your Raspberry Pi and write a little program.

Comparisons

To test to see whether two values are the same, we use ==. This is called a *comparison operator*. The comparison operators we can use are shown in Table 3-2.

Comparison	Description	Example
==	Equals	`total == 11`
!=	Not equals	`total != 11`
>	Greater than	`total > 10`
<	Less than	`total < 3`
>=	Greater than or equal to	`total >= 11`
<=	Less than or equal to	`total <= 2`

Table 3-2 *Comparison Operators*

You can do some experimenting with these comparison operators in the Python Shell. Here's an example:

```
>>> 10 > 9
True
```

In this case, we have basically said to Python, "Is 10 greater than 9?" Python has replied, "True." Now let's ask Python whether 10 is less than 9:

```
>>> 10 < 9
False
```

Being Logical

You cannot fault the logic. When Python tells us "True" or "False," it is not just displaying a message to us. `True` and `False` are special values called *logical values*. Any condition we use with an `if` statement will be turned into a logical value by Python when it is deciding whether or not to perform the next line.

These logical values can be combined rather like the way you perform arithmetic operations like plus and minus. It does not make sense to add `True` and `True`, but it does make sense sometimes to say `True AND True`.

As an example, if we wanted to display a message every time the total throw of our dice was between 5 and 9, we could write something like this:

```
if total >= 5 and total <= 9:
        print('not bad')
```

As well as and, we can use `or`. We can also use `not` to turn `True` into `False`, and vice versa, as shown here:

```
>>> not True
False
```

Thus, another way of saying the same thing would be to write the following:

```
if not (total < 5 or total > 9):
        print('not bad')
```

Exercise

Try incorporating the preceding test into the dice program. While you are at it, add two more if statements: one that prints "Good Throw!" if the throw is higher than 10 and one that prints "Unlucky!" if the throw is less than 4. Try your program out. If you get stuck, you can look at the solution in the file 3_3_double_dice_solution.py.

Else

In the preceding example, you will see that some of the possible throws can be followed by more than one message. Any of the if lines could print an extra message if the condition is true. Sometimes you want a slightly different type of logic, so that if the condition is true, you do one thing and otherwise you do another. In Python, you use else to accomplish this:

```
>>> a = 7
>>> if a > 7:
        print('a is big')
else:
        print('a is small')

a is small
>>>
```

In this case, only one of the two messages will ever be printed.

Another variation on this is elif, which is short for *else if*. Thus, we could expand the previous example so that there are three mutually exclusive clauses, like this:

```
>>> a = 7
>>> if a > 9:
        print('a is very big')
elif a > 7:
        print('a is fairly big')
else:
        print('a is small')

a is small
>>>
```

While

Another command for looping is while, which works a little differently than for. The command while looks a bit like an if command in that it is immediately followed by a condition. In this case, the condition is for staying in the loop. In other words, the code inside the loop will be executed until the condition is no longer true. This means that you have to be careful to ensure that the condition will at some point be false; otherwise, the loop will continue forever and your program will appear to have hung.

To illustrate the use of while, the dice program has been modified so that it just keeps on rolling until a double 6 is rolled:

```
#3_4_double_dice_while
import random
throw_1 = random.randint(1, 6)
throw_2 = random.randint(1, 6)
while not (throw_1 == 6 and throw_2 == 6):
        total = throw_1 + throw_2
        print(total)
        throw_1 = random.randint(1, 6)
        throw_2 = random.randint(1, 6)
print('Double Six thrown!')
```

This program will work. Try it out. However, it is a little bigger than it should be. We are having to repeat the following lines twice—once before the loop starts and once inside the loop:

```
throw_1 = random.randint(1, 6)
throw_2 = random.randint(1, 6)
```

A well-known principle in programming is DRY (Don't Repeat Yourself). Although it's not a concern in a little program like this, as programs get more complex, you need to avoid the situation where the same code is used in more than one place, which makes the programs difficult to maintain.

We can use the command break to shorten the code and make it a bit "drier." When Python encounters the command break, it breaks out of the loop. Here is the program again, this time using break:

```
#3_5_double_dice_while_break
import random
while True:
        throw_1 = random.randint(1, 6)
        throw_2 = random.randint(1, 6)
        total = throw_1 + throw_2
        print(total)
        if throw_1 == 6 and throw_2 == 6:
                break
print('Double Six thrown!')
```

The condition for staying in the loop is permanently set to True. The loop will continue until it gets to break, which will only happen after throwing a double 6.

Summary

You should now be happy to play with IDLE, trying things out in the Python Shell. I strongly recommend that you try altering some of the examples from this chapter, changing the code and seeing how that affects what the programs do.

In the next chapter, we will move on past numbers to look at some of the other types of data you can work with in Python.

4

Strings, Lists, and Dictionaries

This chapter could have had "and Functions" added to its title, but it was already long enough. In this chapter, you will first explore and play with the various ways of representing data and adding some structure to your programs in Python. You will then put everything you learned together into the simple game of Hangman, where you have to guess a word chosen at random by asking whether that word contains a particular letter.

The chapter ends with a reference section that tells you all you need to know about the most useful built-in functions for math, strings, lists, and dictionaries.

String Theory

No, this is not the Physics kind of String Theory. In programming, a *string* is a sequence of characters you use in your program. In Python, to make a variable that contains a string, you can just use the regular = operator to make the assignment, but rather than assigning the variable a number value, you assign it a string value by enclosing that value in single quotes, like this:

```
>>> book_name = 'Programming Raspberry Pi'
```

If you want to see the contents of a variable, you can do so either by entering just the variable name into the Python Shell or by using the `print` command, just as we did with variables that contain a number:

```
>>> book_name
'Programming Raspberry Pi'
>>> print(book_name)
Programming Raspberry Pi
```

There is a subtle difference between the results of each of these methods. If you just enter the variable name, Python puts single quotes around it so that you can tell it is a string. On the other hand, when you use `print`, Python just prints the value.

NOTE   *You can also use double quotes to define a string, but the convention is to use single quotes unless you have a reason for using double quotes (for example, if the string you want to create has an apostrophe in it).*

You can find out how many characters a string has in it by doing this:

```
>>> len(book_name)
24
```

You can find the character at a particular place in the string like so:

```
>>> book_name[1]
'r'
```

Two things to notice here: first, the use of square brackets rather than the parentheses that are used for parameters and, second, that the positions start at 0 and not 1. To find the first letter of the string, you need to do the following:

```
>>> book_name[0]
'P'
```

If you put a number in that is too big for the length of the string, you will see this:

```
>>> book_name[100]
Traceback (most recent call last):
  File "<stdin>", line 1, in <module>
IndexError: string index out of range
>>>
```

This is an error, and it's Python's way of telling us that we have done something wrong. More specifically, the "string index out of range" part of the message tells us that we have tried to access something that we can't. In this case, that's element 100 of a string that is only 24 characters long.

You can chop lumps out of a big string into a smaller string, like this:

```
>>> book_name[0:11]
'Programming'
```

The first number within the brackets is the starting position for the string we want to chop out, and the second number is not, as you might expect, the position of the last character you want, but rather the last character plus 1.

As an experiment, try and chop out the word *raspberry* from the title. If you do not specify the second number, it will default to the end of the string:

```
>>> book_name[12:]
'Raspberry Pi'
```

Similarly, if you do not specify the first number, it defaults to 0.

Finally, you can also join strings together by using + operator. Here's an example:

```
>>> book_name + ' by Simon Monk'
'Programming Raspberry Pi by Simon Monk'
```

Lists

Earlier in the book when you were experimenting with numbers, a variable could only hold a single number. Sometimes, however, it is useful for a variable to hold a list of numbers or strings, or a mixture of both—or even a list of lists. Figure 4-1 will help you to visualize what is going on when a variable is a list.

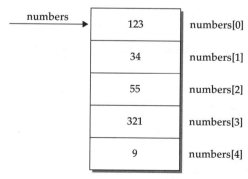

Figure 4-1 *An array*

Lists behave rather like strings. After all, a string is a list of characters. The following example shows you how to make a list. Notice how `len` works on lists as well as strings:

```
>>> numbers = [123, 34, 55, 321, 9]
>>> len(numbers)
5
```

Square brackets are used to indicate a list, and just like with strings we can use square brackets to find an individual element of a list or to make a shorter list from a bigger one:

```
>>> numbers[0]
123
>>> numbers[1:3]
[34, 55]
```

What's more, you can use = to assign a new value to one of the items in the list, like this:

```
>>> numbers[0] = 1
>>> numbers
[1, 34, 55, 321, 9]
```

This changes the first element of the list (element 0) from 123 to just 1.

As with strings, you can join lists together using the + operator:

```
>>> more_numbers = [5, 66, 44]
>>> numbers + more_numbers
[1, 34, 55, 321, 9, 5, 66, 44]
```

If you want to sort the list, you can do this:

```
>>> numbers.sort()
>>> numbers
[1, 9, 34, 55, 321]
```

To remove an item from a list, you use the command pop, as shown next. If you do not specify an argument to pop, it will just remove the last element of the list and return it.

```
>>> numbers
[1, 9, 34, 55, 321]
>>> numbers.pop()
321
>>> numbers
[1, 9, 34, 55]
```

If you specify a number as the argument to pop, that is the position of the element to be removed. Here's an example:

```
>>> numbers
[1, 9, 34, 55]
>>> numbers.pop(1)
9
>>> numbers
[1, 34, 55]
```

As well as removing items from a list, you can also insert an item into the list at a particular position. The function insert takes two arguments. The first is the position before which to insert, and the second argument is the item to insert.

```
>>> numbers
[1, 34, 55]
>>> numbers.insert(1, 66)
>>> numbers
[1, 66, 34, 55]
```

When you want to find out how long a list is, you use len(numbers), but when you want to sort the list or "pop" an element off the list, you put a dot after the variable containing the list and then issue the command, like this:

```
numbers.sort()
```

These two different styles are a result of something called *object orientation*, which we will discuss in the next chapter.

Lists can be made into quite complex structures that contain other lists and a mixture of different types—numbers, strings, and logical values. Figure 4-2 shows the list structure that results from the following line of code:

```
>>> big_list = [123, 'hello', ['inner list', 2, True]]
>>> big_list
[123, 'hello', ['inner list', 2, True]]
```

You can combine what you know about lists with for loops and write a short program that creates a list and then prints out each element of the list on a separate line:

```
#4_1_list_and_for
list = [1, 'one', 2, True]
for item in list:
        print(item)
```

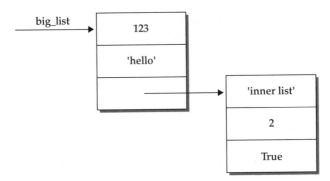

Figure 4-2 *A complex list*

Here's the output of this program:

```
1
one
2
True
```

Functions

When you are writing small programs like the ones we have been writing so far, they only really perform one function, so there is little need to break them up. It is fairly easy to see what they are trying to achieve. As programs get larger, however, things get more complicated and it becomes necessary to break up your programs into units called *functions*. When we get even further into programming, we will look at better ways still of structuring our programs using classes and modules.

Many of the things I have been referring to as *commands* are actually functions that are built into Python. Examples of this are range and print.

The biggest problem in software development of any sort is managing complexity. The best programmers write software that is easy to look at and understand and requires very little in the way of extra explanation. Functions are a key tool in creating easy-to-understand programs that can be changed without difficulty or risk of the whole thing falling into a crumpled mess.

A function is a little like a program within a program. We can use it to wrap up a sequence of commands we want to do. A function that we define can be called from anywhere in our program and will contain its own variables and its own list of commands. When the commands have been run, we are returned to just after wherever it was in the code we called the function in the first place.

As an example, let's create a function that simply takes a string as an argument and adds the word *please* to the end of it. Load the following file—or even better, type it in to a new Editor window—and then run it to see what happens:

```
#4_2_polite_function
def make_polite(sentence):
        polite_sentence = sentence + ' please'
        return polite_sentence

print(make_polite('Pass the salt'))
```

The function starts with the keyword `def`. This is followed by the name of the function, which follows the same naming conventions as variables. After that come the parameters inside parentheses and separated by commas if there are more than one. The first line must end with a colon.

Inside the function, we are using a new variable called `polite_sentence` that takes the parameter passed into the function and adds " please" to it (including the leading space). This variable can only be used from inside the function.

The last line of the function is a `return` command. This specifies what value the function should give back to the code that called it. This is just like trigonometric functions such as `sin`, where you pass in an angle and get back a number. In this case, what is returned is the value in the variable `polite_sentence`.

To use the function, we just specify its name and supply it with the appropriate arguments. A return value is not mandatory, and some functions will just do something rather than calculate something. For example, we could write a rather pointless function that prints "Hello" a specified number of times:

```
#4_3_hello_n
def say_hello(n):
        for x in range(0, n):
                print('Hello')

say_hello(5)
```

This covers the basics of what we will need to do to write our game of Hangman. Although you'll need to learn some other things, we can come back to these later.

Hangman

Hangman is a word-guessing game, usually played with pen and paper. One player chooses a word and draws a dash for each letter of the word, and the other player has to guess the word. They guess a letter at a time. If the letter guessed is not in the word, they lose a life and part of the hangman's scaffold is drawn. If the letter is in the word, all occurrences of the letter are shown by replacing the dashes with the letters.

We are going to let Python think of a word and we will have to guess what it is. Rather than draw a scaffold, Python is just going to tell us how many lives we have left.

You are going to start with how to give Python a list of words to chose from. This sounds like a job for a list of strings:

```
words = ['chicken', 'dog', 'cat', 'mouse', 'frog']
```

The next thing the program needs to do is to pick one of those words at random. We can write a function that does that and test it on its own:

```
#4_4_hangman_words
import random

words = ['chicken', 'dog', 'cat', 'mouse', 'frog']

def pick_a_word():
        word_position = random.randint(0, len(words) - 1)
        return words[word_position]

print(pick_a_word())
```

Run this program a few times to check that it is picking different words from the list.

This is a good start, but it needs to fit into the structure of the game. The next thing to do is to define a new variable called lives_remaining. This will be an integer that we can start off at 14 and decrease by 1 every time a wrong guess is made. This type of variable is called a *global* variable, because unlike variables defined in functions, we can access it from anywhere in the program.

As well as a new variable, we are also going to write a function called `play` that controls the game. We know what `play` should do, we just don't have all the details yet. Therefore, we can write the function `play` and make up other functions that it will call, such as `get_guess` and `process_guess`, as well as use the function `pick_a_word` we've just written. Here it is:

```
def play():
        word = pick_a_word()
        while True:
                guess = get_guess(word)
                if process_guess(guess, word):
                        print('You win! Well Done!')
                        break
                if lives_remaining == 0:
                        print('You are Hung!')
                        print('The word was: ' + word)
                        break
```

A game of Hangman first involves picking a word. There is then a loop that continues until either the word is guessed (`process_guess` returns `True`) or `lives_remaining` has been reduced to zero. Each time around the loop, we ask the user for another guess.

We cannot run this at the moment because the functions `get_guess` and `process_guess` don't exist yet. However, we can write what are called *stubs* for them that will at least let us try out our `play` function. Stubs are just versions of functions that don't do much; they are stand-ins for when the full versions of the functions are written.

```
def get_guess(word):
        return 'a'

def process_guess(guess, word):
        global lives_remaining
        lives_remaining = lives_remaining - 1
        return False
```

The stub for `get_guess` just simulates the player always guessing the letter *a*, and the stub for `process_guess` always assumes that the player guessed wrong and, thus, decreases `lives_remaining` by 1 and returns `False` to indicate that they didn't win.

The stub for `process_guess` is a bit more complicated. The first line tells Python that the `lives_remaining` variable is the global variable of that name. Without that line, Python assumes that it is a new variable local to the function. The stub then reduces the lives remaining by 1 and returns False to indicate that the user has not won yet. Eventually, we will put in checks to see if the player has guessed all the letters or the whole word.

Open the file 4_5_hangman_play.py and run it. You will get a result similar to this:

```
You are Hung!
The word was: dog
```

What happened here is that we have whizzed through all 14 guesses very quickly, and Python has told us what the word was and that we have lost.

All we need to do to complete the program is to replace the stub functions with real functions, starting with get_guess, shown here:

```
def get_guess(word):
        print_word_with_blanks(word)
        print('Lives Remaining: ' + str(lives_remaining))
        guess = input(' Guess a letter or whole word?')
        return guess
```

The first thing get_guess does is to tell the player the current state of their efforts at guessing (something like "c--c--n") using the function print_word. This is going to be another stub function for now. The player is then told how many lives they have left. Note that because we want to append a number (lives_remaining) after the string Lives Remaining:, the number variable must be converted into a string using the built-in str function.

The built-in function input prints the message in its parameter as a prompt and then returns anything that the user types. Note that in Python 2, the input function was called raw_input. Therefore, if you decide to use Python 2, change this function to raw_input.

Finally, the get_guess function returns whatever the user has typed.

The stub function print_word just reminds us that we have something else to write later:

```
def print_word_with_blanks(word):
        print('print_word_with_blanks:not done yet')
```

Open the file 4_6_hangman_get_guess.py and run it. You will get a result similar to this:

```
not done yet
Lives Remaining: 14
 Guess a letter or whole word?x
not done yet
Lives Remaining: 13
 Guess a letter or whole word?y
not done yet
Lives Remaining: 12
 Guess a letter or whole word?
```

Enter guesses until all your lives are gone to verify that you get the "losing" message.

Next, we can create the proper version of print_word. This function needs to display something like "c--c--n," so it needs to know which letters the player has guessed and which they haven't. To do this, it uses a new global variable (this time a string) that contains all the guessed letters. Every time a letter is guessed, it gets added to this string:

```
guessed_letters = ''
```

Here is the function itself:

```
def print_word_with_blanks(word):
        display_word = ''
        for letter in word:
                if guessed_letters.find(letter) > -1:
                    # letter found
                    display_word = display_word + letter
                else:
                    # letter not found
                    display_word = display_word + '-'
        print display_word
```

This function starts with an empty string and then steps through each letter in the word. If the letter is one of the letters that the player has already guessed, it is added to display_word; otherwise, a hyphen (-) is added. The built-in function find is used to check whether the letter is in the guessed_letters. The find function returns -1 if the letter is not there; otherwise, it returns the position of the letter. All we really care about is whether or not it is there, so we just check that the result is greater than -1. Finally, the word is printed out.

Currently, every time `process_guess` is called, it doesn't do anything with the guess because it's still a stub. We can make it a bit less of a stub by having it add the guessed letter to `guessed_letters`, like so:

```
def process_guess(guess, word):
        global lives_remaining
        global guessed_letters
        lives_remaining = lives_remaining - 1
        guessed_letters = guessed_letters + guess
        return False
```

Open the file 4_7_hangman_print_word.py and run it. You will get a result something like this:

```
-------
Lives Remaining: 14
 Guess a letter or whole word?c
c--c---
Lives Remaining: 13
 Guess a letter or whole word?h
ch-c---
Lives Remaining: 12
 Guess a letter or whole word?
```

It's starting to look like the proper game now. However, there is still the stub for `process_guess` to fill out. We will do that next:

```
def process_guess(guess, word):
        if len(guess) > 1:
                return whole_word_guess(guess, word)
        else:
                return single_letter_guess(guess, word)
```

When the player enters a guess, they have two choices: They can either enter a single-letter guess or attempt to guess the whole word. In this method, we just decide which type of guess it is and call either whole_word_guess or single_letter_guess. Because these functions are both pretty straightforward, we will implement them directly rather than as stubs:

```
def single_letter_guess(guess, word):
        global guessed_letters
        global lives_remaining
        if word.find(guess) == -1:
                # word guess was incorrect
                lives_remaining = lives_remaining - 1
        guessed_letters = guessed_letters + guess
```

```
        if all_letters_guessed(word):
                return True

def all_letters_guessed(word):
        for letter in word:
                if guessed_letters.find(letter) == -1:
                        return False
        return True
```

The function whole_word_guess is actually easier than the single_
letter_guess function:

```
def whole_word_guess(guess, word):
        global lives_remaining
        if guess.lower() == word.lower():
                return True
        else:
                lives_remaining = lives_remaining - 1
                return False
```

All we have to do is compare the guess and the actual word to see if they are the same when they are both converted to lowercase. If they are not the same, a life is lost. The function returns True if the guess was correct; otherwise, it returns False.

That's the complete program. Open up 4_8_hangman_full.py in the IDLE Editor and run it. The full listing is shown here for convenience:

```
#04_08_hangman_full
import random

words = ['chicken', 'dog', 'cat', 'mouse', 'frog']
lives_remaining = 14
guessed_letters = ''


def play():
        word = pick_a_word()
        while True:
                guess = get_guess(word)
                if process_guess(guess, word):
                        print('You win! Well Done!')
                        break
                if lives_remaining == 0:
                        print('You are Hung!')
                        print('The word was: ' + word)
                        break
```

```
def pick_a_word():
        word_position = random.randint(0, len(words) - 1)
        return words[word_position]

def get_guess(word):
        print_word_with_blanks(word)
        print('Lives Remaining: ' + str(lives_remaining))
        guess = input(' Guess a letter or whole word?')
        return guess

def print_word_with_blanks(word):
        display_word = ''
        for letter in word:
                if guessed_letters.find(letter) > -1:
                        # letter found
                        display_word = display_word + letter
                else:
                        # letter not found
                        display_word = display_word + '-'
        print(display_word)

def process_guess(guess, word):
        if len(guess) > 1:
                return whole_word_guess(guess, word)
        else:
                return single_letter_guess(guess, word)


def whole_word_guess(guess, word):
        global lives_remaining
        if guess == word:
                return True
        else:
                lives_remaining = lives_remaining - 1
                return False

def single_letter_guess(guess, word):
        global guessed_letters
        global lives_remaining
        if word.find(guess) == -1:
                # letter guess was incorrect
                lives_remaining = lives_remaining - 1
        guessed_letters = guessed_letters + guess
        if all_letters_guessed(word):
                return True
    return False
```

```
def all_letters_guessed(word):
    for letter in word:
        if guessed_letters.find(letter) == -1:
            return False
    return True

play()
```

The game as it stands has a few limitations. First, it is case sensitive, so you have to enter your guesses in lowercase, the same as the words in the `words` array. Second, if you accidentally type **aa** instead of **a** as a guess, it will treat this as a whole-word guess, even though it is too short to be the whole word. The game should probably spot this and only consider guesses the same length as the secret word to be whole-word guesses.

As an exercise, you might like to try and correct these problems. Hint: For the case-sensitivity problem, experiment with the built-in function `lower`. You can look at a corrected version in the file 4_8_hangman_full_solution.py.

Dictionaries

Lists are great when you want to access your data starting at the beginning and working your way through, but they can be slow and inefficient when they get large and you have a lot of data to trawl through (for example, looking for a particular entry). It's a bit like having a book with no index or table of contents. To find what you want, you have to read through the whole thing.

Dictionaries, as you might guess, provide a more efficient means of accessing a data structure when you want to go straight to an item of interest. When you use a dictionary, you associate a value with a key. Whenever you want that value, you ask for it using the key. It's a little bit like how a variable name has a value associated with it; however, the difference is that with a dictionary, the keys and values are created while the program is running.

Let's look at an example:

```
>>> eggs_per_week = {'Penny': 7, 'Amy': 6, 'Bernadette': 0}
>>> eggs_per_week['Penny']
7
>>> eggs_per_week['Penny'] = 5
>>> eggs_per_week
{'Amy': 6, 'Bernadette': 0, 'Penny': 5}
>>>
```

This example is concerned with recording the number of eggs each of my chickens is currently laying. Associated with each chicken's name is a number of eggs per week. When we want to retrieve the value for one of the hens (let's say Penny), we use that name in square brackets instead of the index number that we would use with a list. We can use the same syntax in assignments to change one of the values.

For example, if Bernadette were to a lay an egg, we could update our records by doing this:

```
eggs_per_week['Bernadette'] = 1
```

You may have noticed that when the dictionary is printed, the items in it are not in the same order as we defined them. The dictionary does not keep track of the order in which items were defined. Also note that although we have used a string as the key and a number as the value, the key could be a string, a number, or a tuple (see the next section), but the value could be anything, including a list or another dictionary.

Tuples

On the face of it, tuples look just like lists, but without the square brackets. Therefore, we can define and access a tuple like this:

```
>>> tuple = 1, 2, 3
>>> tuple
(1, 2, 3)
>>> tuple[0]
1
```

However, if we try to change an element of a tuple, we get an error message, like this one:

```
>>> tuple[0] = 6
Traceback (most recent call last):
  File "<stdin>", line 1, in <module>
TypeError: 'tuple' object does not support item assignment
```

The reason for this error message is that tuples are *immutable*, meaning that you cannot change them. Strings and numbers are the same. Although you can change a variable to refer to a different string, number, or tuple, you cannot change the number itself. On the other hand, if the variable references a list, you could alter that list by adding, removing, or changing elements in it.

So, if a tuple is just a list that you cannot do much with, you might be wondering why you would want to use one. The answer is, tuples provide a useful way of creating a temporary collection of items. Python lets you do a couple of next tricks using tuples, as described in the next two subsections.

Multiple Assignment

To assign a value to a variable, you just use = operator, like this:

```
a = 1
```

Python also lets you do multiple assignments in a single line, like this:

```
>>> a, b, c = 1, 2, 3
>>> a
1
>>> b
2
>>> c
3
```

Multiple Return Values

Sometimes in a function, you want to return more than one value at a time. As an example, imagine a function that takes a list of numbers and returns the minimum and the maximum. Here is such an example:

```
#04_09_stats
def stats(numbers):
        numbers.sort()
        return (numbers[0], numbers[-1])

list = [5, 45, 12, 1, 78]
min, max = stats(list)
print(min)
print(max)
```

This method of finding the minimum and maximum is not terribly efficient, but it is a simple example. The list is sorted and then we take the first and last numbers. Note that numbers[-1] returns the last number because when you supply a negative index to an array or string, Python counts backward from the end of the list or string. Therefore, the position -1 indicates the last element, -2 the second to last, and so on.

Exceptions

Python uses exceptions to flag that something has gone wrong in your program. Errors can occur in any number of ways while your program is running. A common way we have already discussed is when you try to access an element of a list or string that is outside of the allowed range. Here's an example:

```
>>> list = [1, 2, 3, 4]
>>> list[4]
Traceback (most recent call last):
  File "<stdin>", line 1, in <module>
IndexError: list index out of range
```

If someone gets an error message like this while they are using your program, they will find it confusing to say the least. Therefore, Python provides a mechanism for intercepting such errors and allowing you to handle them in your own way:

```
try:
        list = [1, 2, 3, 4]
        list[4]
except IndexError:
        print('Oops')
```

We cover exceptions again in the next chapter, where you will learn about the hierarchy of the different types of error that can be caught.

Summary of Functions

This chapter was written to get you up to speed with the most important features of Python as quickly as possible. By necessity, we have glossed over a few things and left a few things out. Therefore, this section provides a reference of some of the key features and functions available for the main types we have discussed. Treat it as a resource you can refer back to as you progress though the book, and be sure to try out some of the functions to see how they work. There is no need to go through everything in this section—just know that it is here when you need it. Remember, the Python Shell is your friend.

For full details of everything in Python, refer to http://docs.python.org/py3k.

Numbers

Table 4-1 shows some of the functions you can use with numbers.

Function	Description	Example
abs(x)	Returns the absolute value (removes the - sign).	>>>abs(-12.3) 12.3
bin(x)	Used to convert to binary string.	>>> bin(23) '0b10111'
complex(r,i)	Creates a complex number with real and imaginary components. Used in science and engineering.	>>> complex(2,3) (2+3j)
hex(x)	Used to convert to hexadecimal string.	>>> hex(255) '0xff'
oct(x)	Used to convert to octal string.	>>> oct(9) '0o11'
round(x, n)	Round x to n decimal places.	>>> round(1.111111, 2) 1.11
math. factorial(n)	Factorial function (as in 4 × 3 × 2 × 1).	>>> math.factorial (4)24
math.log(x)	Natural logarithm.	>>> math.log(10) 2.302585092994046
math.pow(x, y)	Raises x to the power of y (alternatively, use x ** y).	>>> math.pow(2, 8) 256.0
math.sqrt(x)	Square root.	>>> math.sqrt(16) 4.0
math.sin, cos, tan, asin, acos, atan	Trigonometry functions (radians).	>>> math.sin(math.pi / 2) 1.0

Table 4-1 *Number Functions*

Strings

String constants can be enclosed either with single quotes (most common) or with double quotes. Double quotes are useful if you want to include single quotes in the string, like this:

```
s = "Its 3 o'clock"
```

On some occasions you'll want to include special characters such as end-of-lines and tabs into a string. To do this, you use what are called *escape characters*, which begin with a backslash (\) character. Here are the only ones you are likely to need:

- \t   Tab character
- \n   Newline character

Table 4-2 shows some of the functions you can use with strings.

Function	Description	Example
`s.capitalize()`	Capitalizes the first letter and makes the rest lowercase.	`>>> 'aBc'.capitalize()` `'Abc'`
`s.center(width)`	Pads the string with spaces, centering it. An optional extra parameter is used for the fill character.	`>>> 'abc'.center(10, '-')` `'---abc----'`
`s.endswith(str)`	Returns `True` if the end of the string matches.	`>>> 'abcdef'` `.endswith('def')` `True`
`s.find(str)`	Returns the position of a substring. Optional extra arguments for the start and end positions can be used to limit the search.	`>>> 'abcdef'.find('de')` `3`
`s.format(args)`	Formats a string using template markers using { }.	`>>> "Its {0} pm".for-` `mat('12')` `"Its 12 pm"`
`s.isalnum()`	Returns `True` if all the characters in the string are letters or digits.	`>>> '123abc'.isalnum()` `True`
`s.isalpha()`	Returns `True` if all the characters are alphabetic.	`>>> '123abc'.isalpha()` `False`
`s.isspace()`	Returns `True` if the character is a space, tab, or other whitespace character.	`>>> ' \t'.isspace()` `True`
`s.ljust(width)`	Like `center()`, but left-justified.	`>>> 'abc'.ljust(10, '-')` `'abc-------'`
`s.lower()`	Converts a string into lowercase.	`>>> 'AbCdE'.lower()` `'abcde'`
`s.replace(old, new)`	Replaces all occurrences of old with new.	`>>> 'hello world'` `.replace('world',` `'there') 'hello there'`
`s.split()`	Returns a list of all the words in the string, separated by spaces. An optional parameter can be used to indicate a different splitting character. The end of line character (\n) is a popular choice.	`>>> 'abc def'.split()` `['abc', 'def']`
`s.splitlines()`	Splits the string on the newline character.	
`s.strip()`	Removes whitespace from both ends of the string.	`>>> '   a b   '.strip()` `'a b'`
`s.upper()`	Refer to `lower()`, earlier in this table.	

Table 4-2 *String Functions*

Lists

We have already looked at most of the features of lists. Table 4-3 summarizes these features.

Function	Description	Example

`del(a[i:j])`	Deletes elements from the array, from element i to element j-1.	`>>> a = ['a', 'b', 'c']`
		`>>> del(a[1:2])`
		`>>> a`
		`['a', 'c']`

`a.append(x)`	Appends an element to the end of the list.	`>>> a = ['a', 'b', 'c']`
		`>>> a.append('d')`
		`>>> a`
		`['a', 'b', 'c', 'd']`

`a.count(x)`	Counts the occurrences of a particular element.	`>>> a = ['a', 'b', 'a']`
		`>>> a.count('a')`
		`2`

`a.index(x)`	Returns the index position of the first occurrence of x in a. Optional parameters can be used for the start and end index.	`>>> a = ['a', 'b', 'c']`
		`>>> a.index('b')`
		`1`

`a.insert (i, x)`	Inserts x at position i in the list.	`>>> a = ['a', 'c']`
		`>>> a.insert(1, 'b')`
		`>>> a`
		`['a', 'b', 'c']`

`a.pop()`	Returns the last element of the list and removes it. An optional parameter lets you specify another index position for the removal.	`>>> ['a', 'b', 'c']`
		`>>> a.pop(1)`
		`'b'`
		`>>> a`
		`['a', 'c']`

`a.remove(x)`	Removes the element specified.	`>>> a = ['a', 'b', 'c']`
		`>>> a.remove('c')`
		`>>> a`
		`['a', 'b']`

`a.reverse()`	Reverses the list.	`>>> a = ['a', 'b', 'c']`
		`>>> a.reverse()`
		`>>> a`
		`['c', 'b', 'a']`

`a.sort()`	Sorts the list. Advanced options are available when sorting lists of objects. See the next chapter for details.	

Table 4-3 *List Functions*

Dictionaries

Table 4-4 details a few things about dictionaries that you should know.

Function	Description	Example

`len(d)`	Returns the number of items in the dictionary.	```>>> d = {'a':1, 'b':2}```
		```>>> len(d)```
		```2```

`del(d[key])`	Deletes an item from the dictionary.	```>>> d = {'a':1, 'b':2}```
		```>>> del(d['a'])```
		```>>> d```
		```{'b': 2}```

`key in d`	Returns `True` if the dictionary (d) contains the key.	```>>> d = {'a':1, 'b':2}```
		```>>> 'a' in d```
		```True```

`d.clear()`	Removes all items from the dictionary.	```>>> d = {'a':1, 'b':2}```
		```>>> d.clear()```
		```>>> d```
		```{}```

`get(key, default)`	Returns the value for the key, or default if the key is not there.	```>>> d = {'a':1, 'b':2}```
		```>>> d.get('c', 'c')```
		```'c'```

Table 4-4   *Dictionary Functions*

Type Conversions

We have already discussed the situation where we want to convert a number into a string so that we can append it to another string. Python contains some built-in functions for converting items of one type to another, as detailed in Table 4-5.

Function	Description	Example

`float(x)`	Converts x to a floating-point number.	```>>> float('12.34')```
		```12.34```
		```>>> float(12)```
		```12.0```

`int(x)`	Optional argument used to specify the number base.	```>>> int(12.34)```
		```12```
		```>>> int('FF', 16)```
		```255```

`list(x)`	Converts x to a list. This is also a handy way to get a list of dictionaries keys.	```>>> list('abc')```
		```['a', 'b', 'c']```
		```>>> d = {'a':1, 'b':2}```
		```>>> list(d)```
		```['a', 'b']```

Table 4-5   *Type Conversions*

Summary

Many things in Python you will discover gradually. Therefore, do not despair at the thought of learning all these commands. Doing so is really not necessary because you can always search for Python commands or look them up.

In the next chapter, we take the next step and see how Python manages object orientation.

5

Modules, Classes, and Methods

In this chapter, we discuss how to make and use our own modules, like the random module we used in Chapter 3. We also discuss how Python implements object orientation, which allows programs to be structured into classes, each responsible for its own behavior. This helps to keep a check on the complexity of our programs and generally makes them easier to manage. The main mechanisms for doing this are classes and methods. You have already used built-in classes and methods in earlier chapters without necessarily knowing it.

Modules

Most computer languages have a concept like modules that allows you to create a group of functions that are in a convenient form for others to use—or even for yourself to use on different projects.

Python does this grouping of functions in a very simple and elegant way. Essentially, any file with Python code in it can be thought of as a module with the same name as the file. However, before we get into writing our own modules, let's look at how we use the modules already installed with Python.

Using Modules

When we used the `random` module previously, we did something like this:

```
>>> import random
>>> random.randint(1, 6)
6
```

The first thing we do here is tell Python that we want to use the `random` module by using the `import` command. Somewhere in the Python installation is a file called random.py that contains a `randint` function as well as some other functions.

With so many modules available to us, there is a real danger that different modules might have functions with the same name. In such a case, how would Python know which one to use? Fortunately, we do not have to worry about this happening because we have imported the module, and none of the functions in the module are visible unless we prepend the module name and then a dot onto the front of the function name. Try omitting the module name, like this:

```
>>> import random
>>> randint(1, 6)
Traceback (most recent call last):
  File "<stdin>", line 1, in <module>
NameError: name 'randint' is not defined
```

Having to put the module name in front of every call to a function that's used a lot can get tedious. Fortunately, we can make this a little easier by adding to the `import` command as follows:

```
>>> import random as r
>>> r.randint(1,6)
2
```

This gives the module a local name within our program of just `r` rather than `random`, which saves us a bit of typing.

If you are certain a function you want to use from a library is not going to conflict with anything in your program, you can take things a stage further, as follows:

```
>>> from random import randint
>>> randint(1, 6)
5
```

To go even further, you can import everything from the module in one fell swoop. Unless you know exactly what is in the module, however, this is not normally a good idea, but you can do it. Here's how:

```
>>> from random import *
>>> randint(1, 6)
2
```

In this case, the asterisk (*) means "everything."

Useful Python Libraries

So far we have used the `random` module, but other modules are included in Python. These modules are often called Python's *standard library*. There are too many of these modules to list in full. However, you can always find a complete list of Python modules at http://docs.python.org/release/3.1.5 /library/index.html. Here are some of the most useful modules you should take a look at:

- **string** String utilities
- **datetime** For manipulating dates and times
- **math** Math functions (sin, cos, and so on)
- **pickle** For saving and restoring data structures on file (see Chapter 6)
- **urllib.request** For reading web pages (see Chapter 6)
- **tkinter** For creating graphical user interfaces (see Chapter 7)

Installing New Modules

In addition to the standard library modules, thousands of modules have been contributed by the Python community. One very popular module is pygame, which we will use in Chapter 8. It's often available as a binary package, so you can install it by typing something like this:

```
sudo apt-get install python-pygame
```

For many modules, however, this is not the case, and you have to go through a bit more effort to install them.

Any module good enough to use will be packaged in the standard Python way. This means that to install it, you need to download a compressed file containing a directory for the module. Let's use the RPi.GPIO module we will use in Chapter 11 as an example. To install this module, you first go to the module's website, find the Downloads section, and then fetch the archive file. This is shown in Figure 5-1. The next step is to save the file to a convenient directory (for this example, use the Python directory we created in Chapter 3).

Once the file has been saved, open LXTerminal and use cd to get to the Python directory, like so:

```
pi@raspberrypi:~/Python$ ls
RPi.GPIO-0.3.1a.tar.gz
```

Next, you need to extract the directory from the archive by entering the following command:

```
pi@raspberrypi:~/Python$ tar -xzf RPi.GPIO-0.3.1a.tar.gz
pi@raspberrypi:~/Python$ ls
RPi.GPIO-0.3.1a   RPi.GPIO-0.3.1a.tar.gz
```

Now that you have a new folder for the module, you need to "cd" into it and then run the install command. However, it is always worth checking the instructions first to see if there's anything else you need to do. To see the instructions, type **more INSTALL.txt**.

Good thing you checked! The instructions state that you need to do the following:

```
sudo apt-get install python3-dev
```

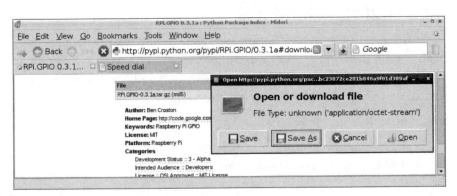

Figure 5-1 *Downloading the RPi.GPIO module*

Finally, you are ready to run the module installer itself:

```
pi@raspberrypi:~/Python$ cd RPi.GPIO-0.3.1a
pi@raspberrypi:~/Python/RPi.GPIO-0.3.1a$ sudo python3
setup.py install
```

Once the module is installed, you will be able to import it from the Python Shell.

Object Orientation

Object orientation has much in common with modules. It shares the same goals of trying to group related items together so that they are easy to maintain and find. As the name suggests, object orientation is about objects. We have been unobtrusively using objects already. A string is an object, for example. Thus, when we type

```
>>> 'abc'.upper()
```

We are telling the string 'abc' that we want a copy of it, but in uppercase. In object-oriented terms, abc is an *instance* of the built-in class str and upper is a *method* on the class str.

We can actually find out the class of an object, as shown here (note double underscores before and after the word class):

```
>>> 'abc'.__class__
<class 'str'>
>>> [1].__class__
<class 'list'>
>>> 12.34.__class__
<class 'float'>
```

Defining Classes

That's enough of other people's classes; let's make some of our own. We are going to start by creating a class that does the job of converting measurements from one unit to another by multiplying a value by a scale factor.

We will give the class the catchy name ScaleConverter. Here is the listing for the whole class, plus a few lines of code to test it:

```
#05_01_converter
class ScaleConverter:
```

```
def __init__(self, units_from, units_to, factor):
    self.units_from = units_from
    self.units_to = units_to
    self.factor = factor

def description(self):
    return 'Convert ' + self.units_from + ' to ' + self.units_to

def convert(self, value):
    return value * self.factor
c1 = ScaleConverter('inches', 'mm', 25)
print(c1.description())
print('converting 2 inches')
print(str(c1.convert(2)) + c1.units_to)
```

This requires some explanation. The first line is fairly obvious: It states that we are beginning the definition of a class called `ScaleConverter`. The colon (:) on the end indicates that all that follows is part of the class definition until we get back to an indent level of the left margin again.

Inside the `ScaleConverter`, we can see what look like three function definitions. These functions belong to the class; they cannot be used except via an instance of the class. These kinds of functions that belong to a class are called *methods*.

The first method, __init__, looks a bit strange—its name has two underscore characters on either side. When Python is creating a new instance of a class, it automatically calls the method __init__. The number of parameters that __init__ should have depends on how many parameters are supplied when an instance of the class is made. To unravel that, we need to look at this line at the end of the file:

```
c1 = ScaleConverter('inches', 'mm', 25)
```

This line creates a new instance of the `ScaleConverter`, specifying what the units being converted from and to are, as well as the scaling factor. The __init__ method must have all these parameters, but it must also have a parameter called `self` as the first parameter:

```
def __init__(self, units_from, units_to, factor):
```

The parameter `self` refers to the object itself. Now, looking at the body of the __init__ method, we see some assignments:

```
self.units_from = units_from
self.units_to = units_to
self.factor = factor
```

Each of these assignments creates a variable that belongs to the object and has its initial value set from the parameters passed in to __init__.

To recap, when we create a new `ScaleConverter` by typing something like

```
c1 = ScaleConverter('inches', 'mm', 25)
```

Python creates a new instance of `ScaleConverter` and assigns the values `'inches'`, `'mm'`, and 25 to its three variables: `self.units_from`, `self.units_to`, and `self.factor`.

The term *encapsulation* is often used in discussions of classes. It is the job of a class to encapsulate everything to do with the class. That means storing data (like the three variables) and things that you might want to do with the data in the form of the `description` and `convert` methods.

The first of these (`description`) takes the information that the Converter knows about its units and creates a string that describes it. As with __init__, all methods must have a first parameter of `self`. The method will probably need it to access the data of the class to which it belongs.

Try it yourself by running program 05_01_converter.py and then typing the following in the Python Shell:

```
>>> silly_converter = ScaleConverter('apples', 'grapes', 74)
>>> silly_converter.description()
'Convert apples to grapes'
```

The `convert` method has two parameters: the mandatory `self` parameter and a parameter called `value`. The method simply returns the result of multiplying the value passed in by `self.scale`:

```
>>> silly_converter.convert(3)
222
```

Inheritance

The `ScaleConverter` class is okay for units of length and things like that; however, it would not work for something like converting temperature from degrees Celsius (C) to degrees Fahrenheit (F). The formula for this is $F = C * 1.8 + 32$. There is both a scale factor (1.8) and an offset (32).

Let's create a class called `ScaleAndOffsetConverter` that is just like `ScaleConverter`, but with a `factor` as well as an `offset`. One way to do this would simply be to copy the whole of the code for `ScaleConverter` and change it a bit by adding the extra variable. It might, in fact, look something like this:

```
#05_02_converter_offset_bad
class ScaleAndOffsetConverter:

    def __init__(self, units_from, units_to, factor, offset):
        self.units_from = units_from
        self.units_to = units_to
        self.factor = factor
        self.offset = offset

    def description(self):
        return 'Convert ' + self.units_from + ' to ' + self.units_to

    def convert(self, value):
        return value * self.factor + self.offset

c2 = ScaleAndOffsetConverter('C', 'F', 1.8, 32)
print(c2.description())
print('converting 20C')
print(str(c2.convert(20)) + c2.units_to)
```

Assuming we want both types of converters in the program we are writing, then this is a bad way of doing it. It's bad because we are repeating code. The `description` method is actually identical, and `__init__` is almost the same. A much better way is to use something called *inheritance*.

The idea behind inheritance in classes is that when you want a specialized version of a class that already exists, you inherit all the parent class's variables and methods and just add new ones or override the ones that are different. Figure 5-2 shows a class diagram for the two classes, indicating how

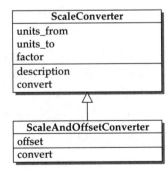

Figure 5-2   *An example of using inheritance*

ScaleAndOffsetConverter inherits from ScaleConverter, adds a new variable (offset), and overrides the method convert (because it will work a bit differently).

Here is the class definition for ScaleAndOffsetConverter using inheritance:

```
class ScaleAndOffsetConverter(ScaleConverter):

    def __init__(self, units_from, units_to, factor, offset):
        ScaleConverter.__init__(self, units_from, units_to, factor)
        self.offset = offset

    def convert(self, value):
        return value * self.factor + self.offset
```

The first thing to notice is that the class definition for ScaleAndOffsetConverter has ScaleConverter in parentheses immediately after it. That is how you specify the parent class for a class.

The __init__ method for the new "subclass" of ScaleConverter first invokes the __init__ method of ScaleConverter before defining the new variable offset. The convert method will override the convert method in the parent class because we need to add on the offset for this kind of converter. You can run and experiment with the two classes together by running 05_03_converters_final.py:

```
>>> c1 = ScaleConverter('inches', 'mm', 25)
>>> print(c1.description())
Convert inches to mm
>>> print('converting 2 inches')
converting 2 inches
>>> print(str(c1.convert(2)) + c1.units_to)
50mm
>>> c2 = ScaleAndOffsetConverter('C', 'F', 1.8, 32)
>>> print(c2.description())
Convert C to F
>>> print('converting 20C')
converting 20C
>>> print(str(c2.convert(20)) + c2.units_to)
68.0F
```

It's a simple matter to convert these two classes into a module that we can use in other programs. In fact, we will use this module in Chapter 7, where we attach a graphical user interface to it.

To turn this file into a module, we should first take the test code off the end of it and then give the file a more sensible name. Let's call it converters.py. You will find this file in the downloads for this book. The module must be in the same directory as any program that wants to use it.

To use the module now, just do this:

```
>>> import converters
>>> c1 = converters.ScaleConverter('inches', 'mm', 25)
>>> print(c1.description())
Convert inches to mm
>>> print('converting 2 inches')
converting 2 inches
>>> print(str(c1.convert(2)) + c1.units_to)
50mm
```

Summary

Lots of modules are available for Python, and some are specifically for the Raspberry Pi, such as the RPi.GPIO library for controlling the GPIO pins. As you work through this book, you will encounter various modules. You will also find that as the programs you write get more complex, the benefits of an object-oriented approach to designing and coding your projects will keep everything more manageable.

In the next chapter, we look at using files and the Internet.

6

Files and the Internet

Python makes it easy for your programs to use files and connect to the Internet. You can read data from files, write data to files, and fetch content from the Internet. You can even check for new mail and tweet—all from your program.

Files

When you run a Python program, any values you have in variables will be lost. Files provide a means of making data more permanent.

Reading Files

Python makes reading the contents of a file extremely easy. As an example, we can convert the Hangman program from Chapter 4 to read the list of words from a file rather than have them fixed in the program.

First of all, start a new file in IDLE and put some words in it, one per line. Then save the file with the name hangman_words.txt in the same directory as the Hangman program from Chapter 4 (04_08_hangman_full.py). Note that in the Save dialog you will have to change the file type to .txt (see Figure 6-1).

Before we modify the Hangman program itself, we can just experiment with reading the file in the Python console. Enter the following into the console:

```
>>> f = open('Python/hangman_words.txt')
```

Note that the Python console has a current directory of /home/pi, so the directory Python (or wherever you saved the file) must be included.

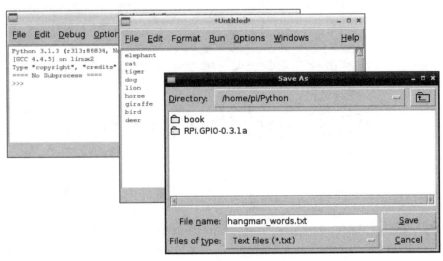

Figure 6-1   *Creating a text file in IDLE*

Next enter the following into the Python console:

```
>>> words = f.read()
>>> words
'elephant\ncat\ntiger\ndog\nlion\nhorse\ngiraffe\nbird\ndeer\n'
>>> words.splitlines()
['elephant', 'cat', 'tiger', 'dog', 'lion', 'horse', 'giraffe'

, 'bird', 'deer']
>>>
```

I told you it was easy! All we need to do to add this file to the Hangman program is replace the line

```
words = ['chicken', 'dog', 'cat', 'mouse', 'frog']
```

with the following lines:

```
f = open('hangman_words.txt')
words = f.read().splitlines()
f.close()
```

The line `f.close()` has been added. You should always call the `close` command when you are done with a file to free up operating system resources. Leaving a file open can lead to problems.

The full program is contained in the file 06_01_hangman_file.py, and a suitable list of animal names can be found in the file hangman_words.txt.

This program does nothing to check that the file exists before trying to read it. So, if there file isn't there, we get an error that looks something like this:

```
Traceback (most recent call last):
  File "06_01_hangman_file.py", line 4, in <module>
    f = open('hangman_words.txt')
IOError: [Errno 2] No such file or directory: 'hangman_words.txt'
```

To make this a bit more user friendly, the file-reading code needs to be inside a try command, like this:

```
try:
        f = open('hangman_words.txt')
        words = f.read().splitlines()
        f.close()
except IOError:
        print("Cannot find file 'hangman_words.txt'")
        exit()
```

Python will try to open the file, but because the file is missing it will not be able to. Therefore, the except part of the program will apply, and the more friendly message will be displayed. Because we cannot do anything without a list of words to guess, there is no point in continuing, so the exit command is used to quit.

In writing the error message, we have repeated the name of the file. Sticking strictly to the Don't Repeat Yourself (DRY) principle, the filename should be put in a variable, as shown next. That way, if we decide to use a different file, we only have to change the code in one place.

```
words_file = 'hangman_words.txt'
try:
        f = open(words_file)
        words = f.read().splitlines()
        f.close()
except IOError:
        print("Cannot find file: " + words_file)
        exit()
```

A modified version of Hangman with this code in it can be found in the file 06_02_hangman_file_try.py.

Reading Big Files

The way we did things in the previous section is fine for a small file containing some words. However, if we were reading a really huge file (say, several megabytes), then two things would happen. First, it would take a significant amount of time for Python to read all the data. Second, because all the data is read at once, at least as much memory as the file size would be used, and for truly enormous files, that might result in Python running out of memory.

If you find yourself in the situation where you are reading a big file, you need to think about how you are going to handle it. For example, if you were searching a file for a particular string, you could just read one line of the file at a time, like this:

```
#06_03_file_readline
words_file = 'hangman_words.txt'
try:
        f = open(words_file)
        line = f.readline()
        while line != '':
                if line == 'elephant\n':
                        print('There is an elephant in the file')
                        break
                line = f.readline()
        f.close()
except IOError:
        print("Cannot find file: " + words_file)
```

When the function `readline` gets to the last line of the file, it returns an empty string (' '). Otherwise, it returns the contents of the line, including the end-of-line character (\n). If it reads a blank line that is actually just a gap between lines and not the end of the file, it will return just the end-of-line character (\n). By the program only reading one line at a time, the memory being used is only ever equivalent to one full line.

If the file is not broken into convenient lines, you can specify an argument in `read` that limits the number of characters read. For example, the following will just read the first 20 characters of a file:

```
>>> f = open('hangman_words.txt')
>>> f.read(20)
'elephant\ncat\ntiger\nd'
>>> f.close()
```

Writing Files

Writing files is almost as simple. When a file is opened, as well as specifying the name of the file to open, you can also specify the mode in which to open the file. The mode is represented by a character, and if no mode is specified it is assumed to be r for read. The modes are as follows:

- **r (read)**.

- **w (write)**   Replaces the contents of any existing file with that name.

- **a (append)**   Appends anything to be written onto the end of an existing file.

- **r+**   Opens the file for both reading and writing (not often used).

To write a file, you open it with a second parameter of ' w ', ' a ', or ' r+ '. Here's an example:

```
>>> f = open('test.txt', 'w')
>>> f.write('This file is not empty')
>>> f.close()
```

The File System

Occasionally, you will need to do some file-system-type operations on files (moving them, copying them, and so on). Python uses Linux to perform these actions, but provides a nice Python-style way of doing them. Many of these functions are in the shutil (shell utility) package. There's a number of subtle variations on the basic copy and move features that deal with file permissions and metadata. In this section, we just deal with the basic operations. You can refer to the official Python documentation for any other functions (http://docs.python.org/release/3.1.5/library).

Here's how to copy a file:

```
>>> import shutil
>>> shutil.copy('test.txt', 'test_copy.txt')
```

To move a file, either to change its name or move it to a different directory:

```
shutil.move('test_copy.txt', 'test_dup.txt')
```

This works on directories as well as files. If you want to copy an entire folder—including all its contents and its content's contents—you can use the function copytree. The rather dangerous function rmtree, on the other hand, will recursively remove a directory and all its contents—exercise extreme caution with this one!

The nicest way of finding out what is in a directory is via *globbing*. The package glob allows you to create a list of files in a directory by specifying a wildcard (*). Here's an example:

```
>>> import glob
glob.glob('*.txt')
['hangman_words.txt', 'test.txt', 'test_dup.txt']
```

If you just want all the files in the folder, you could use this:

```
glob.glob('*')
```

Pickling

Pickling involves saving the contents of a variable to a file in such a way that the file can be later loaded to get the original value back. The most common reason for wanting to do this is to save data between runs of a program. As an example, we can create a complex list containing another list and various other data objects and then pickle it into a file called mylist.pickle, like so:

```
>>> mylist = ['a', 123, [4, 5, True]]
>>> mylist
['a', 123, [4, 5, True]]
>>> import pickle
>>> f = open('mylist.pickle', 'w')
>>> pickle.dump(mylist, f)
>>> f.close()
```

If you find the file and open it in an editor to have a look, you will see something cryptic that looks like this:

```
(lp0
S'a'
p1
aI123
a(lp2
I4
aI5
aI01
aa.
```

That is to be expected; it is text, but it is not meant to be in human-readable form. To reconstruct a pickle file into an object, here is what you do:

```
>>> f = open('mylist.pickle')
>>> other_array = pickle.load(f)
>>> f.close()
>>> other_array
['a', 123, [4, 5, True]]
```

Internet

Most applications use the Internet in one way or another, even if it is just to check whether a new version of the application is available to remind the user about. You interact with a web server by sending HTTP (Hypertext Transfer Protocol) requests to it. The web server then sends a stream of text back as a response. This text will be HTML (Hypertext Markup Language), the language used to create web pages.

Try entering the following code into the Python console.

```
>>> import urllib.request
>>> u = 'http://www.amazon.com/s/ref=nb_sb_noss?field-keywords=raspberry+pi'
>>> f = urllib.request.urlopen(u)
>>> contents = f.read()
... lots of HTML
>>> f.close()
```

Note that you will need to execute the read line as soon as possible after opening the URL. What you have done here is to send a web request to www .amazon.com, asking it to search on "raspberry pi." This has sent back the HTML for Amazon's web page that would display (if you were using a browser) the list of search results.

If you look carefully at the structure of this web page, you can see that you can use it to provide a list of Raspberry Pi–related items found by Amazon. If you scroll around the text, you will find some lines like these:

```
<div class="productTitle"><a href="http://www.amazon
.com/Raspberry-User-Guide

-Gareth-Halfacree/dp/111846446X"> Raspberry Pi User Guide</a> <span

class="ptBrand">by <a href="/Gareth-Halfacree/e
/B0088CA5ZM">Gareth
```

```
Halfacree</a> and Eben Upton</span><span
class="binding"> (<span class
```

```
="format">Paperback</span> - Nov. 13, 2012)</span></div>
```

They key thing here is `<div class="productTitle">`. There is one instance of this before each of the search results. (It helps to have the same web page open in a browser for comparison.) What you want to do is copy out the actual title text. You could do this by finding the position of the text `productTitle`, counting two > characters, and then taking the text from that position until the next < character, like so:

```
#06_04_amazon_scraping
import urllib.request

u = 'http://www.amazon.com/s/ref=nb_sb_noss?field-
keywords=raspberry+pi'
f = urllib.request.urlopen(u)
contents = str(f.read())
f.close()
i = 0
while True:
        i = contents.find('productTitle', i)
        if i == -1:
                break
        # Find the next two '>' after 'productTitle'
        i = contents.find('>', i+1)
        i = contents.find('>', i+1)
        # Find the first '<' after the two '>'
        j = contents.find('<', i+1)
        title = contents[i+2:j]
        print(title)
```

When you run this, you will mostly get a list of products. If you really get into this kind of thing, then search for "Regular Expressions in Python" on the Internet. Regular expressions are almost a language in their own right; they are used for doing complex searches and validations of text. They are not easy to learn or use, but they can simplify tasks like this one.

What we have done here is called *web scraping,* and it is not ideal for a number of reasons. First of all, organizations often do not like people "scraping" their web pages with automated programs. Therefore, you may get a warning or even banned from some sites.

Second, this action is very dependent on the structure of the web page. One tiny change on the website and everything could stop working. A much better approach is to look for an official web service interface to the site. Rather than returning the data as HTML, these services return much more easily processed data, often in XML or JSON format.

If you want to learn more about how to do this kind of thing, search the Internet for "web services in Python."

Summary

This chapter has given you the basics of how to use files and access web pages from Python. There is actually a lot more to Python and the Internet, including accessing e-mail and other Internet protocols. For more information on this, have a look at the Python documentation at http://docs.python .org/release/3.1.5/library/internet.html.

7

Graphical User Interfaces

Everything we have done so far has been text based. In fact, our Hangman game would not have looked out of place on a 1980s home computer. This chapter shows you how to create applications with a proper graphical user interface (GUI).

Tkinter

Tkinter is the Python interface to the Tk GUI system. Tk is not specific to Python; there are interfaces to it from many different languages, and it runs on pretty much any operating system, including Linux. Tkinter comes with Python, so there is no need to install anything. It is also the most commonly used tool for creating a GUI for Python.

Hello World

Tradition dictates that the first program you write with a new language or system should do something trivial, just to show it works! This usually means making the program display a message of "Hello World." As you'll recall, we already did this for Python back in Chapter 3, so I'll make no apologies for starting with this program:

```
#07_01_hello.py

from tkinter import *
root = Tk()
Label(root, text='Hello World').pack()
root.mainloop()
```

Figure 7-1 *Hello World in Tkinter*

Figure 7-1 shows the rather unimpressive application.

You don't need to worry about how all this works. You do, however, need to know that you must assign a variable to the object Tk. Here, we call this variable root, which is a common convention. We then create an instance of the class Label, whose first argument is root. This tells Tkinter that the label belongs to it. The second argument specifies the text to display in the label. Finally, the method pack is called on the label. This tells the label to pack itself into the space available. The method pack controls the layout of the items in the window. Shortly, we will use an alternative type of layout for the components in a grid.

Temperature Converter

To get started with Tkinter, you'll gradually build up a simple application that provides a GUI for temperature conversion (see Figure 7-2). This application will use the converter module we created in Chapter 5 to do the calculation.

Our Hello World application, despite being simple, is not well structured and would not lend itself well to a more complex example. It is normal when building a GUI with Tkinter to use a class to represent each application window. Therefore, our first step is to make a framework in which to slot the

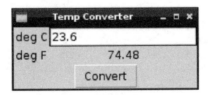

Figure 7-2 *A temperature conversion application*

application, starting with a window with the title "Temp Converter" and a single label:

```
#07_02_temp_framework.py

from tkinter import *

class App:

    def __init__(self, master):
        frame = Frame(master)
        frame.pack()
        Label(frame, text='deg C').grid(row=0, column=0)
        button = Button(frame, text='Convert', command=self.convert)
        button.grid(row=1)

    def convert(self):
        print('Not implemented')

root = Tk()
root.wm_title('Temp Converter')
app = App(root)
root.mainloop()
```

We have added a class to the program called App. It has an __init__ method that is used when a new instance of App is created in the following line:

```
app = App(root)
```

We pass in the Tk root object to __init__ where the user interface is constructed.

As with the Hello World example, we are using a Label, but this time rather than adding the label to the root Tk object, we add the label to a Frame object that contains the label and other items that will eventually make up the window for our application. The structure of the user interface is shown in Figure 7-3. Eventually, it will have all the elements shown.

The frame is "packed" into the root, but this time when we add the label, we use the method grid instead of pack. This allows us to specify a grid layout for the parts of our user interface. The field goes at position 0, 0 of the grid, and the button object that is created on the subsequent line is put on the second row of the grid (row 1). The button definition also specifies a "command" to be run when the button is clicked. At the moment, this is just a stub that prints the message "Not implemented."

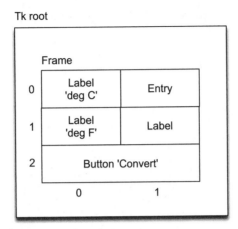

Figure 7-3 *Structure of the user interface*

The function wm_title sets the title of the window. Figure 7-4 shows what the basic user interface looks like at this point.

The next step is to fill in the rest of the user interface. We need an "entry" into which a value for degrees C can be entered and two more labels. We need one permanent label that just reads "deg F" and a label to the right of it where the converted temperature will be displayed.

Tkinter has a special way of linking fields on the user interface with values. Therefore, when we need to get or set the value entered or displayed on a label or entry, we create an instance of a special variable object. This comes in various flavors, and the most common is StringVar. However, because we are entering and displaying numbers, we will use DoubleVar. *Double* means a double-precision floating-point number. This is just like a float, but more precise.

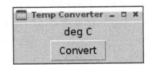

Figure 7-4 *The basic user interface for the Temp Converter application*

After we add in the rest of the user interface controls and the variables to interact with them, the program will look like this:

```
#07_03_temp_ui.py

from tkinter import *

class App:

    def __init__(self, master):
        frame = Frame(master)
        frame.pack()
        Label(frame, text='deg C').grid(row=0, column=0)
        self.c_var = DoubleVar()
        Entry(frame, textvariable=self.c_var).grid(row=0, column=1)
        Label(frame, text='deg F').grid(row=1, column=0)
        self.result_var = DoubleVar()
        Label(frame, textvariable=self.result_var).grid(row=1, column=1)
        button = Button(frame, text='Convert', command=self.convert)
        button.grid(row=2, columnspan=2)

    def convert(self):
        print('Not implemented')

root = Tk()
root.wm_title('Temp Converter')
app = App(root)
root.mainloop()
```

The first DoubleVar (c_var) is assigned to the entry by specifying a textvariable property for it. This means that the entry will display what is in that DoubleVar, and if the value in the DoubleVar is changed, the field display will automatically update to show the new value. Also, when the user types something in the entry field, the value in the DoubleVar will change. Note that a new label of "deg F" has also been added.

The second DoubleVar is linked to another label that will eventually display the result of the calculation. We have added another attribute to the grid command that lays out the button. Because we specify columnspan=2, the button will stretch across both columns.

If you run the program, it will display the final user interface, but when you click the Convert button, the message "Not Implemented" will be written to the Python console.

The last step is to replace the stubbed-out "convert" method with a real method that uses the converters module from Chapter 5. To do this, we need to import the module. In order to reduce how much we need to type, we will import everything, as follows:

```
from converters import *
```

For the sake of efficiency, it is better if we create a single "converter" during __init__ and just use the same one every time the button is clicked. Therefore, we create a variable called self.t_conv to reference the convertor. The convert method then just becomes this:

```
def convert(self):
        c = self.c_var.get()
        self.result_var.set(self.t_conv.convert(c))
```

Here is the full listing of the program:

```
#07_04_temp_final.py

from tkinter import *
from converters import *

class App:

    def __init__(self, master):
        self.t_conv = ScaleAndOffsetConverter('C', 'F', 1.8, 32)
        frame = Frame(master)
        frame.pack()
        Label(frame, text='deg C').grid(row=0, column=0)
        self.c_var = DoubleVar()
        Entry(frame, textvariable=self.c_var).grid(row=0, column=1)
        Label(frame, text='deg F').grid(row=1, column=0)
        self.result_var = DoubleVar()
        Label(frame, textvariable=self.result_var).grid(row=1, column=1)
        button = Button(frame, text='Convert', command=self.convert)
        button.grid(row=2, columnspan=2)

    def convert(self):
        c = self.c_var.get()
        self.result_var.set(self.t_conv.convert(c))

root = Tk()
root.wm_title('Temp Converter')
app = App(root)
root.mainloop()
```

Other GUI Widgets

In the temperature converter, we just used text fields (class Entry) and labels (class Label). As you would expect, you can build lots of other user interface controls into your application. Figure 7-5 shows the main screen of a "kitchen sink" application that illustrates most of the controls you can use in Tkinter. This program is available as 07_05_kitchen_sink.py.

Figure 7-5 *A "kitchen sink" application*

Checkbutton

The Checkbox widget (first column, second row of Figure 7-5) is created like this:

```
Checkbutton(frame, text='Checkbutton')
```

This line of code just creates a Checkbutton with a label next to it. If we have gone to the effort of placing a check box on the window, we'll also want a way of finding out whether or not it is checked.

The way to do this is to use a special "variable" like we did in the temperature converter example. In the following example, we use a StringVar, but if the values of onvalue and offvalue were numbers, we could use an IntVar instead.

```
check_var = StringVar()
check = Checkbutton(frame, text='Checkbutton',
          variable=check_var, onvalue='Y', offvalue='N')
check.grid(row=1, column=0)
```

Listbox

To display a list of items from which one or multiple items can be selected, a Listbox is used (refer to the center of Figure 7-5). Here's an example:

```
listbox = Listbox(frame, height=3, selectmode=BROWSE)
for item in ['red', 'green', 'blue', 'yellow', 'pink']:
        listbox.insert(END, item)
listbox.grid(row=1, column=1)
```

In this case, it just displays a list of colors. Each string has to be added to the list individually. The word END indicates that the item should go at the end of the list.

You can control the way selections are made on the Listbox using the selectmode property, which can be set to one of the following:

- **SINGLE**   Only one selection at a time.
- **BROWSE**   Similar to SINGLE, but allows selection using the mouse. This appears to be indistinguishable from SINGLE in Tkinter on the Pi.
- **MULTIPLE**   SHIFT-click to select more than one row.
- **EXTENDED**   Like MULTIPLE, but also allows the CTRL-SHIFT-click selection of ranges.

Unlike with other widgets that use StringVar or some other type of special variable to get values in and out, to find out which items of the Listbox are selected, you have to ask it using the method curselection. This returns a collection of selection indexes. Thus, if the first, second, and fourth items in the list are selected, you will get a list like this:

```
[0, 1, 3]
```

When selectmode is SINGLE, you still get a list back, but with just one value in it.

Spinbox

Spinboxes provide an alternative way of making a single selection from a list:

```
Spinbox(frame, values=('a','b','c')).grid(row=3)
```

The get method returns the currently displayed item in the Spinbox, not its selection index.

Layouts

Laying out the different parts of your application so that everything looks good, even when you resize the window, is one of the most tricky parts of building a GUI.

You will often find yourself putting one kind of layout inside another. For example, the overall shape of the "kitchen sink" application is a 3×3 grid, but within that grid is another frame for the two radio buttons:

```
radio_frame = Frame(frame)
radio_selection = StringVar()
```

```
b1 = Radiobutton(radio_frame, text='portrait',
    variable=radio_selection, value='P')
b1.pack(side=LEFT)
b2 = Radiobutton(radio_frame, text='landscape',
    variable=radio_selection, value='L')
b2.pack(side=LEFT)
radio_frame.grid(row=1, column=2)
```

This approach is quite common, and it is a good idea to sketch out the layout of your controllers on paper before you start writing the code.

One particular problem you will encounter when creating a GUI is controlling what happens when the window is resized. You will normally want to keep some widgets in the same place and at the same size, while allowing other widgets to expand.

As an example of this, we can build a simple window like the one shown in Figure 7-6, which has a Listbox (on the left) that stays the same size and an expandable message area (on the right) that expands as the window is resized.

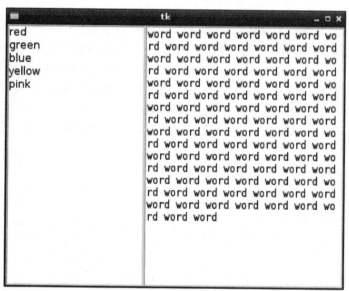

Figure 7-6   *An example of resizing a window*

The code for this is shown here:

```
#07_06_resizing.py

from tkinter import *

class App:

    def __init__(self, master):
        frame = Frame(master)
        frame.pack(fill=BOTH, expand=1)
        #Listbox
        listbox = Listbox(frame)
        for item in ['red', 'green', 'blue', 'yellow', 'pink']:
            listbox.insert(END, item)
        listbox.grid(row=0, column=0, sticky=W+E+N+S)

        #Message
        text = Text(frame, relief=SUNKEN)
        text.grid(row=0, column=1, sticky=W+E+N+S)
        text.insert(END, 'word ' * 100)
        frame.columnconfigure(1, weight=1)
        frame.rowconfigure(0, weight=1)
root = Tk()
app = App(root)
root.geometry("400x300+0+0")
root.mainloop()
```

The key to understanding such layouts is the use of the `sticky` attributes of the components to decide which walls of their grid cell they should stick to. To control which of the columns and rows expand when the window is resized, you use the `columnconfigure` and `rowconfigure` commands. Figure 7-7 shows the arrangement of GUI components that make up this window. The lines indicate where the edge of a user interface item is required to "stick" to its containing wall.

Let's go through the code for this example so that things start to make sense. First, the line

```
frame.pack(fill=BOTH, expand=1)
```

ensures that the frame will fill the enclosing root window so that if the root window changes in size, so will the frame.

Having created the Listbox, we add it to the frame's grid layout using the following line:

```
listbox.grid(row=0, column=0, sticky=W+E+N+S)
```

This specifies that the Listbox should go in position row 0, column 0, but then the `sticky` attribute says that the west, east, north, and south sides of

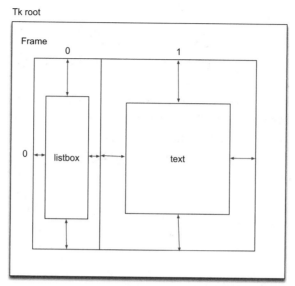

Figure 7-7   *Layout for the resizing window example*

the Listbox should stay connected to the enclosing grid. The constants W, E, N, and S are numeric constants that can be added together in any order. The Text widget is added to the frame's grid in just the same way, and its content is initialized to the word *word* repeated 100 times.

The final part to the puzzle is getting the resizing behavior we want for a text area that expands to the right and a list area that doesn't. To do this, we use the columnconfigure and rowconfigure methods:

```
frame.columnconfigure(1, weight=1)
frame.rowconfigure(0, weight=1)
```

By default, rows and columns do not expand at all when their enclosing user interface element expands. We do not want column 0 to expand, so we can leave that alone. However, we do want column 1 to expand to the right, and we want row 0 (the only row) to be able to expand downward. We do this by giving them a "weight" using the columnconfigure and rowconfigure methods. If, for example, we had multiple columns that we want to expand evenly, we would give them the same weight (typically 1). If, however, we want one of the columns to expand at twice the rate of the other, we would give it twice the weight. In this case, we only have one column and one row that we need expanding, so they can both be given a weight of 1.

Scrollbar

If you shrink down the window for the program 07_06_resizing.py, you will
notice that there's no scrollbar to access text that's hidden. You can still get to
the text, but clearly a scrollbar would help.

Scrollbars are widgets in their own right, and the trick for making them
work with something like a Text, Message, or Listbox widget is to lay them
out next to each other and then link them together.

Figure 7-8 shows a Text widget with a scrollbar.

The code for this is as follows:

```
#07_07_scrolling.py

from tkinter import *

class App:

    def __init__(self, master):
        scrollbar = Scrollbar(master)
        scrollbar.pack(side=RIGHT, fill=Y)
        text = Text(master, yscrollcommand=scrollbar.set)
        text.pack(side=LEFT, fill=BOTH)
        text.insert(END, 'word ' * 1000)
        scrollbar.config(command=text.yview)
```

Figure 7-8   *Scrolling a Text widget*

```
root = Tk()
root.wm_title('Scrolling')
app = App(root)
root.mainloop()
```

In this example, we use the pack layout, positioning the scrollbar on the right and the text area on the left. The fill attribute specifies that the Text widget is allowed to use all free space on *both* the X and Y dimensions.

To link the scrollbar to the Text widget, we set the yscrollcommand property of the Text widget to the set method of the scrollbar. Similarly, the command attribute of the scrollbar is set to text.yview.

Dialogs

It is sometimes useful to pop up a little window with a message and make the user click OK before they can do anything else (see Figure 7-9). These windows are called *modal dialogs,* and Tkinter has a whole range of them in the package tkinter.messagebox.

The following example shows how to display such an alert. As well as showinfo, tkinter.messagebox also has the functions showwarning and showerror that work just the same, but display a different symbol in the window.

```
#07_08_gen_dialogs.py

from tkinter import *
import tkinter.messagebox as mb

class App:

    def __init__(self, master):
        b=Button(master, text='Press Me', command=self.info).pack()
```

Figure 7-9 *An alert dialog*

```
    def info(self):
        mb.showinfo('Information', "Please don't press that button again!")

root = Tk()
app = App(root)
root.mainloop()
```

Other kinds of dialogs can be found in the packages `tkinter.colorchooser` and `tkinter.filedialog`.

Color Chooser

The Color Chooser returns a color as separate RGB components as well as a standard hex color string (see Figure 7-10).

```
#07_09_color_chooser.py

from tkinter import *
import tkinter.colorchooser as cc

class App:

    def __init__(self, master):
        b=Button(master, text='Color..', command=self.ask_color).pack()

    def ask_color(self):
        (rgb, hx) = cc.askcolor()
        print("rgb=" + str(rgb) + " hx=" + hx)

root = Tk()
app = App(root)
root.mainloop()
```

This code returns something like this:

```
rgb=(255.99609375, 92.359375, 116.453125) hx=#ff5c74
```

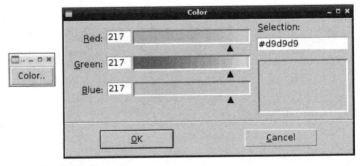

Figure 7-10 *The Color Chooser*

File Chooser

File Choosers can be found in the package `tkinter.filedialog`. These follow exactly the same pattern as the other dialogs we have looked at.

Menus

You can give your applications menus. As an example, we can create a very simple application with an entry field and a couple of menu options (see Figure 7-11).

```
#07_10_menus.py

from tkinter import *

class App:

    def __init__(self, master):
        self.entry_text = StringVar()
        Entry(master, textvariable=self.entry_text).pack()

        menubar = Menu(root)

        filemenu = Menu(menubar, tearoff=0)
        filemenu.add_command(label='Quit', command=exit)
        menubar.add_cascade(label='File', menu=filemenu)

        editmenu = Menu(menubar, tearoff=0)
        editmenu.add_command(label='Fill', command=self.fill)
        menubar.add_cascade(label='Edit', menu=editmenu)

        master.config(menu=menubar)

    def fill(self):
        self.entry_text.set('abc')

root = Tk()
app = App(root)

root.mainloop()
```

Figure 7-11 *Menus*

The first step is to create a root Menu. This is the single object that will contain all the menus (File and Edit, in this case, along with all the menu options).

```
menubar = Menu(root)
```

To create the File menu, with its single option, Quit, we first create another instance of Menu and then add a command for Quit and finally add the File menu to the root Menu:

```
filemenu = Menu(menubar, tearoff=0)
filemenu.add_command(label='Quit', command=exit)
menubar.add_cascade(label='File', menu=filemenu)
```

The Edit menu is created in just the same way. To make the menus appear on the window, we have to use the following command:

```
master.config(menu=menubar)
```

The Canvas

In the next chapter, you'll get a brief introduction to game programming using PyGame. This allows all sorts of nice graphical effects to be achieved. However, if you just need to create simple graphics, such as drawing shapes or plotting line graphs on the screen, you can use Tkinter's Canvas interface instead (see Figure 7-12).

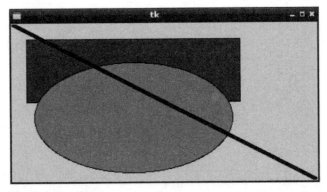

Figure 7-12 *The Canvas widget*

The Canvas is just like any other widget you can add to a window. The following example shows how to draw rectangles, ovals, and lines:

```
#07_11_canvas.py

from tkinter import *

class App:

    def __init__(self, master):
        canvas = Canvas(master, width=400, height=200)
        canvas.pack()
        canvas.create_rectangle(20, 20, 300, 100, fill='blue')
        canvas.create_oval(30, 50, 290, 190, fill='#ff2277')
        canvas.create_line(0, 0, 400, 200, fill='black', width=5)

root = Tk()
app = App(root)
root.mainloop()
```

You can draw arcs, images, polygons, and text in a similar way. Refer to an online Tkinter reference such as http://infohost.nmt.edu/tcc/help/pubs/tkinter/ for more information.

NOTE   *The origin of the coordinates is the top-left corner of the window, and the coordinates are in pixels.*

Summary

In a book this size, it is sometimes only possible to introduce a topic and get you started on the right path. Once you've followed the examples in this chapter, run them, altered them, and analyzed what's going on, you will soon find yourself hungry for more information. You will get past the need for hand-holding and have specific ideas of what you want to write. No book is going to tell you exactly how to build the project you have in your head. This is where the Internet really comes into its own.

Good online references to take what you've learned further can be found here:

- www.pythonware.com/library/tkinter/introduction/
- http://infohost.nmt.edu/tcc/help/pubs/tkinter/

8
Games Programming

Clearly a single chapter is not going to make you an expert in game programming. A number of good books are devoted specifically to game programming in Python, such as *Beginning Game Development with Python and Pygame,* by Will McGugan. This chapter introduces you to a very handy library called pygame and gets you started using it to build a simple game.

What Is Pygame?

Pygame is a library that makes it easier to write games for the Raspberry Pi—or more generally for any computer running Python. The reason why a library is useful is that most games have certain elements in common, and you'll encounter some of the same difficulties when writing them. A library such as pygame takes away some of this pain because someone really good at Python and game programming has created a nice little package to make it easier for us to write games. In particular, pygame helps us in the following ways:

- We can draw graphics that don't flicker.
- We can control the animation so that it runs at the same speed regardless of whether we run it on a Raspberry Pi or a top-of-the-range gaming PC.
- We can catch keyboard and mouse events to control the game play.

The Raspbian Wheezy distribution comes with two versions of Python: Python 2 and Python 3. That is why two shortcuts to IDLE appear on the desktop. So far in this book, we have been using IDLE 3 and thus Python 3.

In Raspbian Wheezy, the Python 3 installation does not include pygame, whereas the Python 2 installation has it preinstalled.

Rather than install pygame into Python 3 (which is a bit involved), we will use Python 2 in this chapter. Don't worry, all the code that we write should still work on Python 3 should you prefer (or find that in a later distribution pygame is there waiting for you). You just have to remember to start IDLE instead of IDLE 3.

Hello Pygame

You may also have a shortcut on your desktop called Python Games. This shortcut runs a launcher program that allows you to run some Python games. However, if you use the File Explorer, you will also find a directory in your root directory called python_games. If you look in here, you will see the .py files for the games, and you can open these files in IDLE to have a look at how others have written their games.

Figure 8-1 shows what a Hello World–type application looks like in pygame, and here is the code listing for it:

```
#08_01_hello_pygame.py

import pygame

pygame.init()
```

Figure 8-1 Hello Pygame

```
screen = pygame.display.set_mode((200, 200))
screen.fill((255, 255, 255))
pygame.display.set_caption('Hello Pygame')

ball = pygame.image.load('raspberry.jpg').convert()
screen.blit(ball, (100, 100))

pygame.display.update()
```

This is a very crude example, and it doesn't have any way of exiting gracefully. Closing the Python console from which this program was launched should kill it after a few seconds.

Looking at the code for this example, you can see that the first thing we do is import pygame. The method init (short for *initialize)* is then run to get pygame set up and ready to use. We then assign a variable called screen using the line

```
screen = pygame.display.set_mode((200, 200))
```

which creates a new window that's 200 by 200 pixels. We then fill it with white (the color 255, 255, 255) on the next line before setting a caption for the window of "Hello Pygame."

Games use graphics, which usually means using images. In this example, we read an image file into pygame:

```
raspberry = pygame.image.load('raspberry.jpg').convert()
```

In this case, the image is a file called raspberry.jpg, which is included along with all the other programs in this book in the programs download section on the book's website. The call to convert() at the end of the line is important because it converts the image into an efficient internal representation that enables it to be drawn very quickly, which is vital when we start to make the image move around the window.

Next, we draw the raspberry image on the screen at coordinates 100, 100 using the blit command. As with the Tkinter canvas you met in the previous chapter, the coordinates start with 0, 0 in the top-left corner of the screen.

Finally, the last command tells pygame to update the display so that we get to see the image.

A Raspberry Game

To show how pygame can be used to make a simple game, we are going to gradually build up a game where we catch falling raspberries with a spoon. The raspberries fall at different speeds and must be caught on the eating end of the spoon before they hit the ground. Figure 8-2 shows the finished game in action. It's crude but functional. Hopefully, you will take this game and improve upon it.

Following the Mouse

Let's start developing the game by creating the main screen with a spoon on it that tracks the movements of the mouse left to right. Load the following program into IDLE:

```
#08_02_rasp_game_mouse

import pygame
from pygame.locals import *
from sys import exit
```

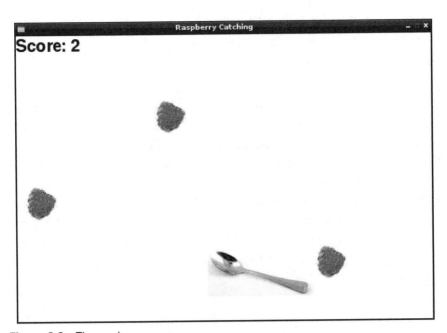

Figure 8-2   *The raspberry game*

```
spoon_x = 300
spoon_y = 300

pygame.init()

screen = pygame.display.set_mode((600, 400))
pygame.display.set_caption('Raspberry Catching')

spoon = pygame.image.load('spoon.jpg').convert()

while True:

    for event in pygame.event.get():
        if event.type == QUIT:
            exit()

    screen.fill((255, 255, 255))
    spoon_x, ignore = pygame.mouse.get_pos()
    screen.blit(spoon, (spoon_x, spoon_y))

    pygame.display.update()
```

The basic structure of our Hello World program is still there, but you have some new things to examine. First of all, there are some more imports. The import for pygame.locals provides us access to useful constants such as QUIT, which we will use to detect when the game is about to exit. The import of exit from sys allows us to quit the program gracefully.

We have added two variables (spoon_x and spoon_y) to hold the position of the spoon. Because the spoon is only going to move left to right, spoon_y will never change.

At the end of the program is a while loop. Each time around the loop, we first check for a QUIT event coming from the pygame system. Events occur every time the player moves the mouse or presses or releases a key. In this case, we are only interested in a QUIT event, which is caused by someone clicking the window close icon in the top-right corner of the game window. We could chose not to exit immediately here, but rather prompt the player to see whether they indeed want to exit. The next line clears the screen by filling it with the color white.

Next comes an assignment in which we set spoon_x to the value of the x position of the mouse. Note that although this is a double assignment, we do not care about the y position of the mouse, so we ignore the second return value by assigning it to a variable called `ignore` that we then ignore. We then draw the spoon on the screen and update the display.

Run the program. You should see the spoon following the mouse.

One Raspberry

The next step in building the game is to add a raspberry. Later on we will expand this so that there are three raspberries falling at a time, but starting with one is easier. The code listing for this can be found in the file 08_03_rasp_game_one.py.

Here are the changes from the previous version:

- Add global variables for the position of the raspberry (`raspberry_x` and `raspberry_y`).
- Load and convert the image raspberry.jpg.
- Separate updating the spoon into its own function.
- Add a new function called `update_raspberry`.
- Update the main loop to use the new functions.

You should already be familiar with the first two items in this list, so let's start with the new functions:

```
def update_spoon():
    global spoon_x
    global spoon_y
    spoon_x, ignore = pygame.mouse.get_pos()
    screen.blit(spoon, (spoon_x, spoon_y))
```

The function `update_spoon` just takes the code we had in the main loop in 08_02_rasp_game_mouse and puts it in a function of its own. This helps to keep the size of the main loop down so that it is easier to tell what's going on.

```
def update_raspberry():
    global raspberry_x
    global raspberry_y
    raspberry_y += 5
    if raspberry_y > spoon_y:
        raspberry_y = 0
        raspberry_x = random.randint(10, screen_width)
    raspberry_x += random.randint(-5, 5)
    if raspberry_x < 10:
        raspberry_x = 10
```

```
if raspberry_x > screen_width - 20:
    raspberry_x = screen_width - 20
screen.blit(raspberry, (raspberry_x, raspberry_y))
```

The function update_raspberry changes the values of raspberry_x and raspberry_y. It adds 5 to the y position to move the raspberry down the screen and moves the x position by a random amount between –5 and +5. This makes the raspberries wobble unpredictably during their descent. However, the raspberries will eventually fall off the bottom of the screen, so once the y position is greater than the position of the spoon, the function moves them back up to the top and to a new random x position.

There is also a danger that the raspberries may disappear off the left or right side of the screen. Therefore, two further tests check that the raspberries aren't too near the edge of the screen, and if they are then they aren't allowed to go any further left or right.

Here's the new main loop that calls these new functions:

```
while True:
    for event in pygame.event.get():
        if event.type == QUIT:
            exit()

    screen.fill((255, 255, 255))
    update_raspberry()
    update_spoon()
    pygame.display.update()
```

Try out 08_03_rasp_game_one. You will see a basically functional program that looks like the game is being played. However, nothing happens when you catch a raspberry.

Catch Detection and Scoring

We are now going to add a message area to display the score (that is, the number of raspberries caught). To do this, we must be able to detect that we have caught a raspberry. The extended program that does this is in the file 08_04_rasp_py_game_scoring.py.

The main changes for this version are two new functions, check_for_catch and display:

```
def check_for_catch():
    global score
```

```
if raspberry_y >= spoon_y and raspberry_x >= spoon_x and \
    raspberry_x < spoon_x + 50:
      score += 1
display("Score: " + str(score))
```

Note that because the condition for the `if` is so long, we use the line-continuation command (\) to break it into two lines.

The function `check_for_catch` adds 1 to the score if the raspberry has fallen as far as the spoon (`raspberry_y >= spoon_y`) and the x position of the raspberry is between the x (left) position of the spoon and the x position of the spoon plus 50 (roughly the width of the business end of the spoon).

Regardless of whether the raspberry is caught, the score is displayed using the `display` function. The `check_for_catch` function is also added into the main loop as one more thing we must do each time around the loop.

The 'display' function is responsible for displaying a message on the screen.

```
def display(message):
    font = pygame.font.Font(None, 36)
    text = font.render(message, 1, (10, 10, 10))
    screen.blit(text, (0, 0))
```

You write text on the screen in pygame by creating a font, in this case, of no specific font family but of a 36-point size and then create a `text` object by rendering the contents of the string `message` onto the font. The value (10, 10, 10) is the text color. The end result contained in the variable `text` can then be blitted onto the screen in the usual way.

Timing

You may have noticed that nothing in this program controls how fast the raspberries fall from the sky. We are lucky in that they fall at the right sort of speed on a Raspberry Pi. However, if we were to run this game on a faster computer, they would probably fly past far too fast to catch.

To manage the speed, pygame has a built-in clock that allows us to slow down our main loop by just the right amount to perform a certain number of refreshes per second. Unfortunately, it can't do anything to speed up our main loop. This clock is very easy to use; you simply put the following line somewhere before the main loop:

```
clock = pygame.time.Clock()
```

This creates an instance of the clock. To achieve the necessary slowing of the main loop, put the following line somewhere in it (usually at the end):

```
clock.tick(30)
```

In this case, we use a value of 30, meaning a frame rate of 30 frames per second. You can put a different value in here, but the human eye (and brain) do not register any improvement in quality above about 30 frames per second.

Lots of Raspberries

Our program is starting to look a little complex. If we were to add the facility for more than one raspberry at this stage, it would become even more difficult to see what is going on. We are therefore going to perform *refactoring*, which means changing a perfectly good program and altering its structure without changing what it actually does or without adding any features. We are going to do this by creating a class called Raspberry to do all the things we need a raspberry to do. This still works with just one raspberry, but will make working with more raspberries easier later. The code listing for this stage can be found in the file 08_05_rasp_game_refactored.py. Here's the class definition:

```
class Raspberry:
    x = 0
    y = 0

    def __init__(self):
        self.x = random.randint(10, screen_width)
        self.y = 0

    def update(self):
        self.y += 5
        if self.y > spoon_y:
            self.y = 0
            self.x = random.randint(10, screen_width)
        self.x += random.randint(-5, 5)
        if self.x < 10:
            self.x = 10
        if self.x > screen_width - 20:
            self.x = screen_width - 20
        screen.blit(raspberry_image, (self.x, self.y))

    def is_caught(self):
        return self.y >= spoon_y and self.x >= spoon_x and \
               self.x < spoon_x + 50
```

The `raspberry_x` and `raspberry_y` variables just become variables of the new `Raspberry` class. Also, when an instance of a raspberry is created, its x position will be set randomly. The old `update_raspberry` function has now become a method on `Raspberry` called just `update`. Similarly, the `check_for_catch` function now asks the raspberry if it has been caught.

Having defined a raspberry class, we create an instance of it like this:

```
r = Raspberry()
```

Thus, when we want to check for a catch, the `check_for_catch` just asks the raspberry like this:

```
def check_for_catch():
    global score
    if r.is_caught():
        score += 1
```

The call to display the score has also been moved out of the `check_for_catch` function and into the main loop. With everything now working just as it did before, it is time to add more raspberries. The final version of the game can be found in the file 08_06_rasp_game_final.py. It is listed here in full:

```
#08_06_rasp_game_final

import pygame
from pygame.locals import *
from sys import exit
import random

score = 0

screen_width = 600
screen_height = 400

spoon_x = 300
spoon_y = screen_height - 100

class Raspberry:
    x = 0
    y = 0
    dy = 0
```

```
    def __init__(self):
        self.x = random.randint(10, screen_width)
        self.y = 0
        self.dy = random.randint(3, 10)

    def update(self):
        self.y += self.dy
        if self.y > spoon_y:
            self.y = 0
            self.x = random.randint(10, screen_width)
        self.x += random.randint(-5, 5)
        if self.x < 10:
            self.x = 10
        if self.x > screen_width - 20:
            self.x = screen_width - 20
        screen.blit(raspberry_image, (self.x, self.y))

    def is_caught(self):
        return self.y >= spoon_y and self.x >= spoon_x

            and self.x < spoon_x + 50

clock = pygame.time.Clock()
rasps = [Raspberry(), Raspberry(), Raspberry()]

pygame.init()

screen = pygame.display.set_mode((screen_width, screen_height))
pygame.display.set_caption('Raspberry Catching')

spoon = pygame.image.load('spoon.jpg').convert()
raspberry_image = pygame.image.load('raspberry.jpg').convert()

def update_spoon():
    global spoon_x
    global spoon_y
    spoon_x, ignore = pygame.mouse.get_pos()
    screen.blit(spoon, (spoon_x, spoon_y))

def check_for_catch():
    global score
    for r in rasps:
        if r.is_caught():
            score += 1

def display(message):
    font = pygame.font.Font(None, 36)
    text = font.render(message, 1, (10, 10, 10))
    screen.blit(text, (0, 0))
```

```
while True:
    for event in pygame.event.get():
        if event.type == QUIT:
            exit()

    screen.fill((255, 255, 255))
    for r in rasps:
        r.update()
    update_spoon()
    check_for_catch()
    display("Score: " + str(score))
    pygame.display.update()
    clock.tick(30)
```

To create multiple raspberries, the single variable `r` has been replaced by a collection called `rasps`:

```
rasps = [Raspberry(), Raspberry(), Raspberry()]
```

This creates three raspberries; we could change it dynamically while the program is running by adding new raspberries to the list (or for that matter removing some).

We now need to make just a couple other changes to deal with more than one raspberry. First of all, in the `check_for_catch` function, we now need to loop over all the raspberries and ask each one whether it has been caught (rather than just the single raspberry). Second, in the main loop, we need to display all the raspberries by looping through them and asking each to update.

Summary

You can learn plenty more about pygame. The official website at www.pygame .org has many resources and sample games that you can play with or modify.

9

Interfacing Hardware

The Raspberry Pi has a double row of pins on one side of it. These pins are called the GPIO connector (General Purpose Input/Output) and allow you to connect electronic hardware to the Pi as an alternative to using the USB port.

The maker and education communities have already started producing expansion and prototyping boards you can attach to your Pi so you can add your own electronics. This includes everything from simple temperature sensors to relays. You can even convert your Raspberry Pi into a controller for a robot.

In this chapter, we explore the various ways of connecting the Pi to electronic devices using the GPIO. We'll use some of the first products that have become available for this purpose. Because this is a fast-moving field, it is fairly certain that new products will have come on the market since this chapter was written; therefore, check the Internet to see what is current. I have tried to choose a representative set of different approaches to interfacing hardware. Therefore, even if the exact same versions are not available, you will at least get a flavor of what is out there and how to use it.

Products to help you attach electronics to your Pi can be categorized as either expansion boards or prototyping tools. Before we look at each of these items, we will look at exactly what the GPIO connector provides us.

GPIO Pin Connections

Figure 9-1 shows the connections available on the Raspberry Pi's GPIO connector. The pins labeled GPIO can all be used as general-purpose input/output pins. In other words, any one of them can first be set to either an input or an

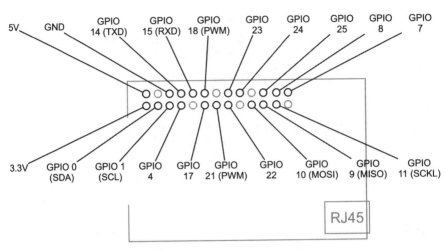

Figure 9-1   *GPIO pin connections*

output. If the pin is set to be an input, you can then test to see whether the pin is set to a "1" (above about 1.7V) or a "0" (below 1.7V). Note that all the GPIO pins are 3.3V pins and connecting them to higher voltages than that could damage your Raspberry Pi.

When set to be an output, the pin can be either 0V or 3.3V (logical 0 or 1). Pins can only supply or sink a small amount of current (assume 5mA to be safe), so they can just light an LED if you use a high value resistor (say, 1kΩ). You will notice that some of the GPIO pins have other letters in parentheses after their names. Those pins have some special purpose. For example, GPIO 0 and 1 have the extra names of SDA and SCL. These are the clock and data lines, respectively, for a serial bus type called I2C that is popular for communicating with peripherals such as temperature sensors, LCD displays, and the like. This I2C bus is used by the Pi Face and Slice of PI/O discussed in the following sections.

GPIO pins 14 and 15 also double as the Rx and Tx (Receive and Transmit) pins for the Raspberry Pi's serial port. Yet another type of serial communication is possible through GPIO 9 to 11 (MISO, MOSI, and SCLK). This type of serial interface is called SPI.

Finally, GPIO 18 and GPIO 21 are labeled PWM, meaning that they are capable of pulse width modulation. This technique allows you to control the power to motors, LEDs etc. by varying the width of pulses generated at a constant rate.

Direct Connection to GPIO Pins

With care, it is possible to attach simple electronics such as LEDs directly to the GPIO pins; however, only do this if you know what you are doing because you could easily damage your Raspberry Pi. In fact, this is more or less what we will be doing in the later section "Prototyping Boards."

Expansion Boards

Expansion boards usually have screw terminals and a certain amount of electronics already built in. This makes them very suitable for educational use as well as for those who do not want to get deeply involved in the electronics side of things. In general, no soldering needs to be done with these kind of boards. They will usually "buffer" all the connections to the Raspberry Pi, which means the Raspberry Pi is protected from anything untoward occurring on the expansion board. For example, a short circuit across an output might damage the expansion board, but no harm will befall your precious Pi.

The sections that follow detail some of the more popular boards, explain their features, and detail how you might go about using them. One such board (the RaspiRobotBoard) will be used to create a simple robot in Chapter 11.

Pi Face

The Pi Face, shown in Figure 9-2, is a board intended primarily for educational use. It was been developed by Manchester University in the UK. As well as providing a useful hardware platform, it also provides an easy-to-use Python library and integration with the Scratch programming environment.

The Pi Face sits on top of the Raspberry Pi and provides convenient screw terminals for connecting devices to it. It does not use the GPIO pins directly, but rather uses as an MCP23S17 port expander chip that it communicates with using the I2C serial interface. This provides eight inputs and eight outputs on the expansion board, but only the two I2C pins on the Raspberry Pi GPIO connector are used. The outputs are provided with further current

Figure 9-2   *The Pi Face expansion board*

amplification using a Darlington driver IC that can supply up to 500mA for each output—more than enough power to directly drive a relay or a 1W high-power LED.

Output devices on the board include two relays that can be used to switch high-load currents. Each relay also has an LED that lights when the relay is activated. There are also two LEDs that can be controlled independently. Four of the inputs have push switches next to them.

The Pi Face has its own Python module that simplifies the use of the board. The following example entered into the Python console shows you how to read digital input 2:

```
>>> import piface.pfio as pfio
>>> pfio.init()
>>> pfio.digital_read(2)
0
```

To turn on digital output 2, you would do the following:

```
>>> pfio.digital_write(2, 1)
```

The LEDs and relays have their own control functions. The following example turns LED 1 on then off again and then turns Relay 1 on:

```
>>> led1 = pfio.LED(1)
>>> led1.turn_on()
>>> led1.turn_off()
>>> relay = pfio.Relay(1)
>>> relay.turn_on()
```

The library must be downloaded and installed. For downloads, documentation, and some sample projects, visit the projects code page at https://github.com/thomasmacpherson/piface. You can also find more information about the project at http://pi.cs.man.ac.uk/interface.htm.

Slice of PI/O

The Slice of PI/O, shown in Figure 9-3, is a small, low-cost board that provides eight buffered inputs and eight buffered outputs using the same MCP23S17 port expander as the Pi Face. It does not, however, have the Darlington driver of the Pi Face and, therefore, cannot drive high-power loads. The maximum load directly from the MCP23S17 is 25mA, which is enough to drive an LED with suitable series resistor, but not enough to drive a relay directly.

The board takes all the I/O pins to edge connectors, and each of the 16 I/O pins can be configured as either an input or output.

Figure 9-3 *The Slice of PI/O*

Here's a list of the key features:

- Sixteen bidirectional buffered I/O connections

- Jumper-selected 3.3V or 5V operation

- Raspberry Pi I2C and SPI serial connections broken out (caution: unbuffered)

- Raspberry Pi GPIO pins 0 to 7 broken out (caution: unbuffered)

At the time of writing, the board is not supplied with any supporting Python module; however, this is likely to change, either through efforts of the supplier or the Raspberry Pi community.

RaspiRobotBoard

I have to declare my personal interest in the RaspiRobotBoard, shown in Figure 9-4, because it is a board I have designed. The focus of this board is firmly on allowing the Raspberry Pi to be used as a robot controller. For this reason, it has a motor controller that allows you to control the direction of two motors (usually attached to wheels).

Another feature that makes it suitable for use as a robot platform is the voltage regulator that powers the Raspberry Pi using any source of power

Figure 9-4 *The RaspiRobotBoard*

between 6V and 9V, such as four AA batteries. The RaspiRobotBoard also has connectors for two different types of serial port, one of which is intended to take an adapter board for an ultrasonic range finder module. The board also has a pair of switch inputs, two LEDs, and another pair of buffered outputs that can be used to drive other LEDs or low-current loads. This board is used in Chapter 11 to build a small roving robot.

Gertboard

The Gertboard is designed by Gert van Loo of Broadcom and therefore is the most official Raspberry Pi expansion board (see Figure 9-5).

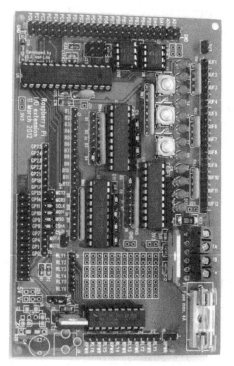

Figure 9-5   *A Gertboard expansion board for the Pi*

The Gertboard is really the kitchen sink of expansion boards. Its key features are as follows:

- Strapping area where GPIO pins can be connected to different modules
- ATmega (like the Arduino) microcontroller
- SPI analog-to-digital and digital-to-analog converters
- Motor controller (like the RaspiRobotBoard)
- 500mA open collector outputs (like the Pi Face)
- Twelve LEDs and three push buttons

Prototyping Boards

Unlike expansion boards, prototyping boards mostly require the use of a soldering iron and a certain amount of electronics expertise. They also connect directly to the Raspberry Pi's main chip, which means that if you get it wrong, you could easily damage your Raspberry Pi. These boards are for the experienced electronics enthusiast—or the very careful or the very reckless (who doesn't mind the possibility of killing their Raspberry Pi).

One of these prototyping boards, the "Cobbler," is not actually a board but rather a connector that allows you to link the GPIO connections to a solderless breadboard where you can add your own electronics. As a contrast to the expansion board approach, we will explore this method further in the next chapter using the Cobbler.

Pi Cobbler

The Pi Cobbler from Adafruit (www.adafruit.com/products/914) comes as a kit that must be soldered together. The soldering is pretty straightforward, and once everything is assembled, you will have a board with 26 pins coming out of the bottom that can be attached to a solderless breadboard (see Figure 9-6). On top of the board is a 26-pin socket to which a 26-way ribbon cable lead (also supplied) can be used to link the Raspberry Pi GPIO connector to the Cobbler.

Pi Plate

The Pi Plate, shown in Figure 9-7, is another product from Adafruit (https://www.adafruit.com/products/801). This is a prototyping board that has a large area in the middle to which you can solder the components for your

Figure 9-6   *The Adafruit Pi Cobbler*

project. Screw terminals are located all around the edge of the board so you can attach leads to external components that won't fit on the board, such as motors and such. In one corner of the board is an area to which a surface mount IC can be soldered. The pins next to it "break out" the difficult-to-use pins of the IC.

Humble Pi

The Humble Pi, shown in Figure 9-8, is quite similar to the Pi Plate, but it lacks the surface mount prototyping area. However, it makes up for this by

Figure 9-7   *The Adafruit Pi Plate*

Figure 9-8   *The Humble Pi*

providing an area where you can add your own voltage regulator and power socket, making it suitable for powering the Pi at 5V from batteries or an external power supply. No voltage regulator or associated capacitors are provided, although Ciseco sells a kit of components for this.

Arduino and the Pi

Although the Raspberry Pi can be used like a microcontroller to drive motors and such, this is not really what it was designed for. As such, the GPIO pins cannot supply much in the way of drive current and are somewhat delicate and intolerant of electrical abuse. This is, after all, the motivation for the expansion boards described in the previous section.

Arduino boards, on the other hand, are much more rugged and designed to be used to control electronic devices (see Figure 9-9). What is more, they have analog inputs that can measure a voltage from, say, a temperature sensor.

Arduino boards are designed to allow communication with a host computer using USB, and there is no reason why this host shouldn't be a Raspberry Pi. This means that the Arduino takes care of all the electronics and the Raspberry Pi sends it commands or listens for incoming requests from the Arduino.

If you have an Arduino, you can try out the following simple example, which allows you to send messages to the Arduino to blink its build-in LED

Figure 9-9 *An Arduino board connected to a Raspberry Pi*

on and off while at the same time receiving incoming messages from the Arduino. Once you can do that, it is easy to adapt either the Arduino sketch or the Python program on the Raspberry Pi to carry out more complex tasks.

This example assumes you are familiar with the Arduino. If you are not, you may want to read some of my other books on the Arduino, including *Programming Arduino: Getting Started with Sketches* and *30 Arduino Projects for the Evil Genius*.

Arduino and Pi Talk

To get the Arduino and Pi to talk, we are going to connect them using a USB port on the Raspberry Pi. Because the Arduino only draws about 50mA and in this case has no external electronics attached to it, it can be powered by the Pi.

The Arduino Software

All you need to do is load the following Arduino sketch onto the Arduino. You will probably want to do this with your regular computer, because at the time of writing, only a very old version of the Arduino software is available for the Raspberry Pi. The following sketch is available in the downloads package and is called PiTest.ino:

```
// Pi and Arduino

const int ledPin = 13;

void setup()
{
  pinMode(ledPin, OUTPUT);
  Serial.begin(9600);
}

void loop()
{
  Serial.println("Hello Pi");
  if (Serial.available())
  {
    flash(Serial.read() - '0');
  }
  delay(1000);
}

void flash(int n)
{
  for (int i = 0; i < n; i++)
  {
    digitalWrite(ledPin, HIGH);
    delay(100);
    digitalWrite(ledPin, LOW);
    delay(100);
  }
}
```

This very simple sketch contains just three functions. The setup function initializes serial communications and sets pin 13 on the LED to be an output. This pin is attached to the LED built into the Arduino. The loop function is invoked repeatedly until the Arduino is powered down. It first sends the message "Hello Pi" to the Raspberry Pi and then checks to see whether there is any incoming communication from the Pi. If there is (it expects a single digit), it flashes the LED on and off that many times using the flash function.

The Raspberry Pi Software

The Python code to talk to the Arduino is even more simple and can just be typed into the Python console. But first, you need to install the PySerial package to allow the communication to take place. This is done in the same way as the other packages we have installed—just fetch the zipped tar file from http://sourceforge.net/projects/pyserial/files/latest/download?source=files.

Next, extract the directory from the archive by entering the following command:

```
tar -xzf pyserial-2.5.tar.gz
```

Now that you have a new folder for the module, just cd into it and then run the install command (first, though, it is worth checking the instructions to see if anything else needs doing beforehand). You are now ready to run the module installer itself, as follows:

```
cd pyserial-2.5
sudo python setup.py install
```

Once it's installed, you will be able to import the module from the Python shell. Now switch from the Linux terminal to a Python console and type the following:

```
import serial
ser = serial.Serial('/dev/ttyACM0', 9600)
```

This opens the USB serial connection with the Arduino at the same baud rate of 9600. Now you need to start a loop listening for messages from the Arduino:

```
while 1 :
    ser.readline()
```

You will need two hit ENTER twice after you type the second line. Messages should now start to appear! You can see in the blue writing where the Arduino is talking to the Pi and then some error trace as you press CTRL-C to interrupt the messages coming from the Arduino.

Now type the following into the Python console:

```
ser.write('5')
```

This should cause the LED to flash five times.

Summary

In this chapter we looked at just some of the wide range of ways of adding electronics to our Raspberry Pi projects. In the next two chapters, we create projects using two different approaches—first using the Adafruit Cobbler and breadboard and then using the RaspiRobotBoard as the basis for a small roving robot.

10

Prototyping Project (Clock)

In this chapter, we will build what can only be seen as a grossly over-engineered LED digital clock. We will be using a Raspberry Pi, Adafruit's Cobbler lead, a breadboard, and a four-digit LED display (see Figure 10-1).

In the first phase of the design, the project will just display the time. However, a second phase extends the project by adding a push button that, when

Figure 10-1 *LED clock using the Raspberry Pi*

pressed, switches the display mode between displaying hours/minutes, seconds, and the date.

What You Need

To build this project, you will need the following parts. Suggested part suppliers are listed, but you can also find these parts elsewhere on the Internet.

Part	Suppliers	Guide Price (in U.S. Dollars)
Raspberry Pi	Farnell, RS Components	$35
Pi Cobbler	Adafruit (Product 914)	$8
Adafruit four-digit seven-segment I2C display	Adafruit (Product 880)	$10
Solderless breadboard	Adafruit (Product 64), Spark-Fun (SKU PRT-00112), Maplin (AG09K)	$5
Assorted jumper wires (male to male) or a solid core wire	Adafruit (Product 758), SparkFun (SKU PRT-08431), Maplin (FS66W)	$8
PCB mount push switch*	Adafruit (Product 367), Spark-Fun (SKU COM-00097), Maplin (KR92A)	$2
* Optional. Only required for Phase Two.		

Hardware Assembly

Both the Pi Cobbler and the display modules from Adafruit come as kits that must be soldered together before they can be used. Both are fairly easy to solder, and detailed step-by-step instructions for building them can be found on the Adafruit website. Each module has pins that just push into the holes on the breadboard.

The display has just four pins (VCC, GND, SDA, and SCL) when it is plugged into the breadboard; align it so that the VCC pin is on row 1 of the breadboard.

The Cobbler has 26 pins, but we will only be using a few of them. It should be inserted at the other end of the breadboard, or at least far enough away so that none of the pins overlap with the same rows as the display. The Cobbler socket has a cutout on one side to ensure that the ribbon cable can only be inserted one way. This cutout should be toward the top of the breadboard, as shown in Figure 10-2.

Figure 10–2   *Breadboard layout*

Underneath the holes of the solderless breadboard are strips of connectors, linking the five holes of a particular row together. Note that because the board is on its side, the rows actually run vertically in Figure 10-2.

Figure 10-2 shows the solderless breadboard with the four pins of the display at one end of the breadboard and the Cobbler at the other. When you're following the instructions in this chapter, it will help if you insert your modules the same way as Figure 10-2 shows.

NOTE   *It is much easier to attach the jumper wires to the breadboard without the ribbon cable attached to the Cobbler.*

The connections that need to be made are listed here:

Suggested Lead Color	From	To
Black	Cobbler GND	Display GND (second pin from left)
Red	Cobbler 5V0	Display VCC (leftmost pin)
Orange	Cobbler SDA0	Display SDA (third pin from left)
Yellow	Cobbler SCL0	Display SCL (rightmost pin)

The color scheme shown in this table is only a suggestion; however, it is common to use red for a positive supply and black or blue for the ground connection.

CAUTION   *In this project, we are connecting a 5V display module to the Raspberry Pi, which generally uses 3.3V. We can only safely do this because the display module used here only acts as a "slave" device and hence only listens on the SDA and SCL lines. Other I2C devices may act as a master device, and if they are 5V, there is a good chance this could damage your Pi. Therefore, before you connect any I2C device to your Raspberry Pi, make sure you understand what you are doing.*

We can now link the Cobbler to the Raspberry Pi using the ribbon cable supplied with the Cobbler. This should be done with the Raspberry Pi powered down. The cable will only fit one way into the Cobbler, but no such protection is provided on the Raspberry Pi. Therefore, make sure the red line on the cable is to the outside of the Raspberry Pi, as shown in Figure 10-1.

Turn on the Raspberry Pi. If the usual LEDs do not light, turn it off immediately and check all the wiring.

Software

Everything is connected, and the Raspberry Pi has booted up. However, the display is still blank because we have not yet written any software to use it. We are going to start with a simple clock that just displays the Raspberry Pi's system time. The Raspberry Pi does not have a real-time clock to tell it the time. However, it will automatically pick up the time from a network time server if it is connected to the Internet.

The Raspberry Pi displays the time in the bottom-right corner of the screen. If the Pi is not connected to the Internet, you can set the time manually using the following command:

```
sudo date -s "Aug 24 12:15"
```

However, you will have to do this every time you reboot. Therefore, it is far better to have your Raspberry Pi connected to the Internet.

If you are using the network time, you may find that the minutes are correct but that the hour is wrong. This probably means that your Raspberry Pi does now know which time zone it is in. This can be fixed by using the following command, which opens up a window where you can select your continent and then the city for the time zone you require:

```
sudo dpkg-reconfigure tzdata
```

At the time of writing, in order to use the I2C bus that the display uses, the Raspbian Wheezy distribution requires that you issue a few commands to make the I2C bus accessible to the Python program we are going to write. It is likely that later releases of Raspbian (and other distributions) will have the port already configured so that the following commands are not necessary. However, for the moment, here is what you need to do:

```
sudo apt-get install python-smbus
sudo modprobe i2c-dev
sudo modprobe i2c-bcm2708
```

NOTE   *You may find that you have to issue the last two of these commands each time you reboot the Raspberry Pi.*

So now that the Raspberry Pi knows the correct time and the I2C bus is available, we can write a Python program that sends the time to the display. To help simplify this process, I have produced a Python library module specifically for this kind of display. It can be downloaded from http://code .google.com/p/i2c7segment/downloads/list.

As with other modules you have installed, you need to fetch the file, extract it into some convenient location (using `tar -xzf`), and then issue the following command to install it under Python 2:

```
sudo python setup.py install
```

The actual clock program itself is contained in the file bundle that accompanies this book (see www.raspberrypibook.com); it is called 10_01_clock.py and is listed here:

```
import i2c7segment as display
import time

disp = display.Adafruit7Segment()

while True:
    h = time.localtime().tm_hour
    m = time.localtime().tm_min
    disp.print_int(h * 100 + m)
    disp.draw_colon(True)
    disp.write_display()
    time.sleep(0.5)
    disp.draw_colon(False)
```

```
disp.write_display()
time.sleep(0.5)
```

The program is nice and simple. The loop continues forever, getting the hour and minute and showing them in the correct places on the display by multiplying the hour by 100 to shift it into the leftmost digits and then adding the minutes that will appear on the right.

The i2c7segment library does most of the work for us. This library is used by first setting what is to be displayed using `print_int` or `draw_colon` and then using `write_display` to update what is displayed.

The colon is made to flash by turning it on, waiting half a second, and then turning it off again. Access to the I2C port is only available to super-users, so you need to run the command as a super-user by entering the following:

```
sudo python 10_01_clock.py
```

If everything is working okay, your display should show the time.

Phase Two

Having got the basic display working, let's expand both the hardware and software by adding a button that changes the mode of the display, cycling between the time in hours and minutes, the seconds, and the date. Figure 10-3 shows the breadboard with the switch added as well as two new patch wires. Note that we are just adding to the layout of the first phase by adding the button; nothing else is changed.

NOTE   *Shut down and power off your Pi before you start making changes on the breadboard.*

The button has four leads and must be placed in the right position; otherwise, the switch will appear to be closed all the time. The leads should emerge from the sides facing the top and bottom of Figure 10-3. Don't worry if you have the switch positioned in the wrong way—it will not damage anything, but the display will continuously change mode without the button being pressed.

Two new wires are needed to connect the switch. One goes from one lead of the switch (refer to Figure 10-3) to the GND connection of the display. The other lead goes to the connection labeled #17 on the Cobbler. The effect is that whenever the button on the switch is pressed, the Raspberry Pi's GPIO 17 pin will be connected to ground.

Figure 10-3   *Adding a button to the design*

You can find the updated software in the file 10_02_fancy_clock.py and listed here:

```
import i2c7segment as display
import time
import RPi.GPIO as io

switch_pin = 17
io.setmode(io.BCM)
io.setup(switch_pin, io.IN, pull_up_down=io.PUD_UP)
disp = display.Adafruit7Segment()

time_mode, seconds_mode, date_mode = range(3)
disp_mode = time_mode

def display_time():
    h = time.localtime().tm_hour
    m = time.localtime().tm_min
    disp.print_int(h * 100 + m)
    disp.draw_colon(True)
    disp.write_display()
    time.sleep(0.5)
    disp.draw_colon(False)
    disp.write_display()
    time.sleep(0.5)
```

```
def disply_date():
    d = time.localtime().tm_mday
    m = time.localtime().tm_mon
    #disp.print_int(d * 100 + m)  # World format
    disp.print_int(m * 100 + d)   # US format
    disp.draw_colon(True)
    disp.write_display()
    time.sleep(0.5)

def display_seconds():
    s = time.localtime().tm_sec
    disp.print_str('----')
    disp.print_int(s)
    disp.draw_colon(True)
    disp.write_display()
    time.sleep(0.5)

while True:
    key_pressed = not io.input(switch_pin)
    if key_pressed:
        disp_mode = disp_mode + 1
        if disp_mode > date_mode:
            disp_mode = time_mode
    if disp_mode == time_mode:
        display_time()
    elif disp_mode == seconds_mode:
        display_seconds()
    elif disp_mode == date_mode:
        disply_date()
```

The first thing to notice is that because we need access to GPIO pin 17 to see whether the button is pressed, we need to use the RPi.GPIO library. We used this as an example of installing a module back in Chapter 5. Therefore, if you have not installed RPi.GPIO, refer back to Chapter 5 and do so now.

We set the switch pin to be an input using the following command:

```
io.setup(switch_pin, io.IN, pull_up_down=io.PUD_UP)
```

This command also turns on an internal pull-up resistor that ensures the input is always at 3.3V (high) unless the switch is pressed to override it and pull it low.

Most of what was in the loop has been separated into a function called `display_time`. Also, two new functions have been added: `display_seconds` and `display_date`. These are fairly self-explanatory.

One point of interest is that `display_date` displays the date in U.S. format. If you want to change this to the international format, where the day of the month comes before the month, change the line that starts with `disp.print_int` appropriately (refer to the comments in the code).

To keep track of which mode we are in, we have added some new variables in the following lines:

```
time_mode, seconds_mode, date_mode = range(3)
disp_mode = time_mode
```

The first of these lines gives each of the three variables a different number. The second line sets the `disp_mode` variable to the value of `time_mode`, which we use later in the main loop.

The main loop has been changed to determine whether the button is pressed. If it is, then 1 is added to `disp_mode` to cycle the display mode. If the display mode has reached the end, it is set back to `time_mode`.

Finally, the `if` blocks that follow select the appropriate display function, depending on the mode, and then call it.

Summary

This project's hardware can quite easily be adapted to other uses. You could, for example, present all sorts of things on the display by modifying the program. Here are some ideas:

- Your current Internet bandwidth (speed)
- The number of e-mails in your inbox
- A countdown of the days remaining in the year
- The number of visitors to a website

In the next chapter, we build another hardware project—this time a roving robot—using the Raspberry Pi as its brain.

11

The RaspiRobot

In this chapter, you will learn how to use the Raspberry Pi as the brain for a simple robot rover, shown in Figure 11-1. The Pi will take commands from a wireless USB keyboard and control the power to motors attached to a robot chassis kit. The robot will also (optionally) have an ultrasonic range finder that tells it how far away obstacles are as well as an LCD screen that displays information from the range finder.

Like the project in the previous chapter, this project is split into two phases. In the first phase, we create a basic rover that you can drive with a wireless keyboard; in the second phase, we add the screen and range finder.

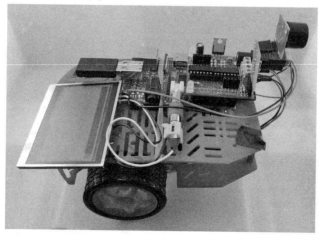

Figure 11-1 *The RaspiRobot*

WARNING   *If batteries are attached to the RaspiRobotBoard, they will supply power to the Raspberry Pi. Do not, under any circumstances, power your Raspberry Pi from its power adaptor and the RaspiRobotBoard at the same time. You can leave the RaspiRobotBoard attached to your Raspberry Pi, but do not attach the motors or batteries to it.*

What You Need

To build this project, you need the following parts. Suggested part suppliers are listed, but you can find other suppliers on the Internet.

Part	Suppliers	Guide Price (in U.S. Dollars)
Raspberry Pi	Farnell, RS Components, CPC, Newark	$35
RaspiRobotBoard	www.raspirobot.com	$TBA
Range finder serial adapter *	www.raspirobot.com	$5
Maxbotix LV-EZ1 serial range finder *	SparkFun (SEN-00639), Adafruit (Product 172)	$25
3.5-inch LCD screen *	Adafruit (Product 913)	$45
Male-to-male RCA adaptor for screen *	Adafruit (Product 951)	$2
Magician Chassis	SparkFun (ROB-10825)	$15
2.1mm power-to-screw terminal adaptor (male) +	Adafruit (Product 369), SparkFun (PRT-10288)	$2
Six AA battery holder	Adafruit (Product 248), + Newark (63J6606), Maplins (HQ01B)	$5
PP3-style battery clip +	RadioShack (270-324), Maplins (NE19V)	$2
Six AA batteries (rechargeable or alkaline)		
Wireless USB keyboard	Computer store or supermarket	$10
* Phase 2 only.		
+ This is a different battery box design than the one I used. It terminates in a 2.1 mm power plug, so the PP3 battery clip is not required if you use this battery box. If you only intend to build the first phase, and you are using the Adafruit battery box, you do not need either the 2.1 mm power-to-screw terminal adaptor (male) or the PP3-style battery clip.		

Phase 1: A Basic Rover

Figure 11-2 shows the basic rover. The basis for this rover is the Magician Chassis kit. This useful kit is composed of a plastic chassis, gear motors, wheels, and all the nuts and bolts to assemble the chassis. It also includes a battery box for four AA batteries, but in this project, that box will be replaced by one that takes six AA batteries.

Hardware Assembly

This project is assembled from a number of different kits of parts. If you search around, you may find already-assembled options when buying the RaspiRobotBoard and the range finder serial adapter, which means the entire project can be built without any soldering (or, in fact, any tools more difficult to use than a screwdriver).

Step 1: Assemble the Chassis

The Magician Chassis comes as a kit of parts that must be assembled. Included in the kit are detailed assembly instructions. When assembling the chassis, you need to replace the supplied battery box (four AA cells) with your six-AA-cell version (see Figure 11-3). If your battery box is the kind that

Figure 11-2   *The basic rover*

Figure 11-3   *Replacing the battery box*

holds two rows of three cells, you will find that the top plate of the Magician Chassis can hold it in place. In fact, it will be quite a tight fit and spring out a little; therefore, you will not need to fit the middle strut.

If your battery box has the cells all in a single row, you will probably need to use the screws that came with the Magician Chassis for its battery box to fix your battery holder securely onto the chassis.

Attach the battery clip to the battery box and the trailing leads from the battery clip to the screw terminal in order to power plug adapter. Be very careful to get the polarity correct (red to plus)!

Before you attach the top surface of the Magician Chassis, slip a rubber band over the top surface. This will be used to hold the Raspberry Pi in place (refer to Figure 11-2).

Step 2: Assemble the RaspiRobotBoard

At the time of writing, it was not clear whether the RaspiRobotBoard will be available already assembled or in kit form only. If it is available in kit form, you need to follow the instructions that accompany it to build the board. Once assembled, the board should look like the one shown in Figure 11-4.

Note that these instructions are for Version 1 of the board. The position of the connectors may change in later versions. See the book's website (www .raspberrypibook.com) for more information. All the connections we are interested in are on the right side of Figure 11-4. At the top is the power socket, and beneath that are the screw terminals for the left and right motors.

Figure 11-4 *The RaspiRobotBoard*

Step 3: Install the Software on the Raspberry Pi

To control the robot, we are going to write a program in Python that detects key presses and use them to control the power to the robot's motors on the robot. To do this, we will use pygame, which provides a useful way of finding out whether or not keys are pressed.

It will be much easier to set the program up before we attach the Raspberry Pi to the chassis. Therefore, attach the RaspiRobotBoard to the Raspberry Pi, but leave the motors and battery disconnected and power up the Pi from your normal USB power supply.

The RaspiRobotBoard has its own Python library, but also relies on some other libraries that must be installed. First of all, it requires the RPi.GPIO library that you first met in Chapter 5 and then again in Chapter 10. If you have not already done so, install the RPi.GPIO library. You will also need to install the PySerial library. See the instructions for this in the Arduino section toward the end of Chapter 9.

The RaspiRobotBoard library can be installed from the following website: http://code.google.com/p/raspirobotboard/downloads/list

Installation is the same as for any other Python package. Because we are using Python 2 in this project, the library should be installed using the command on the following page.

```
tar -xzf raspirobotboard-1.0.tar.gz
cd raspirobotboard-1.0
sudo python setup.py install
```

The actual Python program for this version of the robot is contained in the file 11_01_rover_basic.py, which must be run as super user. Therefore, just to try things out (still with the motors disconnected), run the program by changing to the "code" directory and entering the following in the terminal:

```
sudo python 11_01_rover_basic.py
```

A blank pygame window should appear and the two LEDs go out. We can test the program without the motors because the program sets the LEDs as well as controls the motors. Thus, if you press the UP ARROW key, both LEDs should light once more. Press the SPACEBAR to turn them off again. Then try the LEFT and RIGHT ARROW keys; an LED should light that corresponds to the key you pressed.

We are not going to have leads trailing from our robot to a monitor and mouse, so we need to arrange for this program to automatically start when our Raspberry Pi has finished booting up. To do this, we need to place the file raspirobot_basic.desktop (included in the "code" directory) into a directory named /home/pi/.config/autostart. You can do all this with the File Manager. Just type **/home/pi/.config** in the address bar at the top of the screen. Note that directories that start with a dot are hidden, so you cannot navigate to it in the File Manager simply by clicking.

If there is no directory inside .config called autostart, so create one and copy the file raspirobot_basic.desktop into it. We can make sure our autostart works by rebooting the Pi. If all goes well, the pygame window will appear automatically.

We will return later to look at the code for this project, but for now, let's just get everything working.

Step 4: Connect the Motors

Shut down and disconnect the Raspberry Pi from its power supply. Be sure to put it away so that you do not accidentally apply both it and the battery connection. Put the batteries into the battery holder and fix the top plate of the chassis into place. Cover the metal screws with little patches of insulating tape or Scotch tape to prevent accidental shorts with the Raspberry Pi and then slip the Pi under the elastic band. Next, attach the motors to the terminal block, as shown in Figure 11-5.

Figure 11-5 *Attaching the motors*

Each motor has a red and a black lead. Therefore, find the leads going to the left motor and attach the black lead to the leftmost terminal in Figure 11-5. Attach the red lead from the same motor to the second-from-left terminal block. For the other motor, put the red lead in the third-from-left terminal and the black lead in the remaining screw terminal.

Step 5: Try It Out

That's it. We are ready to go! Attach the USB dongle from the wireless keyboard to the Pi and then attach the plug from the battery lead into the power socket on the RaspiRobotBoard. The LEDs on the Raspberry Pi should flicker as it starts to boot. If this does not happen, immediately disconnect the battery and check your work.

Initially the LEDs on the RaspiRobotBoard should both be lit, but when the Python program starts to run, they should both turn off. Wait another second or two to allow the program to start up properly and then try pressing the ARROW and SPACEBAR keys on your keyboard. Your RaspiRobot should start to move!

About the Software

The software for the first phase is listed here:

```
from raspirobotboard import *
import pygame
import sys
from pygame.locals import *
```

```
rr = RaspiRobot()

pygame.init()
screen = pygame.display.set_mode((640, 480))
pygame.display.set_caption('RaspiRobot')
pygame.mouse.set_visible(0)

while True:
    for event in pygame.event.get():
        if event.type == QUIT:
            sys.exit()
        if event.type == KEYDOWN:
            if event.key == K_UP:
                rr.forward()
                rr.set_led1(True)
                rr.set_led2(True)
            elif event.key == K_DOWN:
                rr.set_led1(True)
                rr.set_led2(True)
                rr.reverse()
            elif event.key == K_RIGHT:
                rr.set_led1(False)
                rr.set_led2(True)
                rr.right()
            elif event.key == K_LEFT:
                rr.set_led1(True)
                rr.set_led2(False)
                rr.left()
            elif event.key == K_SPACE:
                rr.stop()
                rr.set_led1(False)
                rr.set_led2(False)
```

NOTE *If you skipped Chapter 8 on pygame, now might be a good time to read through it.*

The program starts by importing the library modules it needs. It then creates an instance of the class RaspiRobot and assigns it to the variable rr. The main loop first checks for a QUIT event, and if it find one it exists the program. The rest of the loop is concerned with checking all of the keys and issuing the appropriate commands if a key is pressed. For example, if the UP

ARROW key (K_UP) is pressed, the RaspiRobot is sent the command for-ward, which sets the motors to both go forward as well as sets both LEDs on.

Phase 2: Adding a Range Finder and Screen

When you complete Phase 2, your RaspiRobot will look like the one shown earlier in Figure 11-1. Disconnect the battery from the RaspiRobotBoard so that we can start making the necessary changes to complete this phase.

Step 1: Assemble the Range Finder Serial Adapter

The serial range finder module, shown in Figure 11-6, outputs an inverted signal; therefore, a tiny board with a single transistor and resistor on it must be used to invert the signal back to normal. Full instructions for assembling this little adaptor board can be found on the book's website (www.raspberrypibook.com).

The range finder module plugs into the top of the adapter, and the bottom of the adapter fits into the serial socket, as shown in Figure 11-7.

Step 2: Attach the Screen

The LCD screen comes in two parts: the screen itself and the driver board. These are connected together with a rather delicate ribbon cable. I attached the two together with a cushioned self-adhesive pad. Be careful where you attach the screen, and treat the display delicately.

Figure 11-6   *The range finder serial adapter and range finder module*

Figure 11-7   *Assembling the range finder and adapter*

The screen comes with power leads (red and black) as well as two RCA plugs. To neaten things up, I cut off one of the RCA plugs (the one connected to the middle cables in the white connector plug). Refer to Figure 11-8. If this seems too drastic, an alternative is to fasten it somewhere with a cable tie so that it's out of the way.

To the remaining RCA plug I attached the male-to-male RCA adapter. The power leads are then twisted onto the power leads of the same color from the battery clip and inserted into the screw terminals of the plug adapter. If your

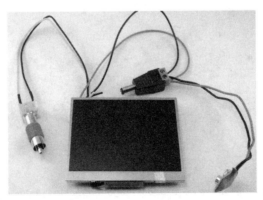

Figure 11-8   *Wiring the display*

battery box is terminated in a plug, you can snip off the plug and strip the insulation of the wires. These wires can then be used as if they were the leads from the battery clip. Either way, the project wiring is summarized in Figure 11-9.

One consequence of this wiring arrangement is that the display will still be connected to the battery, even if you unplug the power on the RaspiRobot-Board. For this reason, use the battery clip on the battery box to turn the robot on and off.

NOTE *More adventurous readers might like to add the luxury of an on/off switch.*

The screen is attached to the chassis by means of adhesive putty. This is not a good permanent solution; some kind of plastic bracket might be better.

Step 3: Update the Software

The updated hardware needs an update to the software to accompany it. The program can be found in the file 11_02_rover_plus.py. You also need to

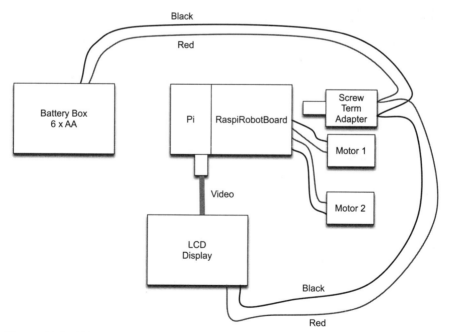

Figure 11-9 *Wiring diagram*

arrange for this program to start rather than the simpler old program. To do this, copy the file raspirobot_plus.desktop into the directory /home/pi/ .config/autostart and remove the raspirobot_basic.desktop file from that folder; otherwise, both programs would start.

Note that because in this phase of the project, the Raspberry Pi has a screen and a keyboard (albeit a very small one), it is possible to make the changes described here, but you'll be using a tiny screen. If this proves too difficult, then as before, disconnect the battery, detach the motors, and power the Raspberry Pi from its USB power supply with its regular monitor, keyboard, and mouse.

Step 4: Run It

That's it! The project is ready to run. As always, if the LEDs on the Raspberry Pi don't come on straight away, disconnect the batteries and look for the problem. The Raspberry Pi is pretty power hungry for a battery-powered device. The screen also uses quite a lot of power. Therefore, to avoid too much recharging of batteries, you should disconnect them when not in use.

Revised Software

The new program is bigger than the old one, so it is not listed here in full. You can open it up in IDLE to take a look. The main differences are, as you would expect, the distance sensing and the display. The function get_range is shown here:

```
def get_range():
    try:
        dist = rr.get_range_inch()
    except:
        dist = 0
    return dist
```

This function is a very thin wrapper around a call to get_range_inch in the RaspiRobot module. The exception handling is added because if the range finder does not work for any reason (say, it isn't plugged in), exceptions will be raised. This function just intercepts any such exceptions and returns a distance of 0 if that happens.

The update_display function first gets the distance and then displays it along with a graphical indication of the closeness of any obstacles, as shown in Figure 11-10.

The code for this is shown here:

```
def update_distance():
    dist = get_range()
    if dist == 0:
        return
    message = 'Distance: ' + str(dist) + ' in'
    text_surface = font.render(message, True, (127, 127, 127))
    screen.fill((255, 255, 255))
    screen.blit(text_surface, (100, 100))

    w = screen.get_width() - 20
    proximity = ((100 - dist) / 100.0) * w
    if proximity < 0:
        proximity = 0
    pygame.draw.rect(screen, (0, 255, 0), Rect((10, 10),(w, 50)))
    pygame.draw.rect(screen, (255, 0, 0), Rect((10, 10),(proximity, 50)))
    pygame.display.update()
```

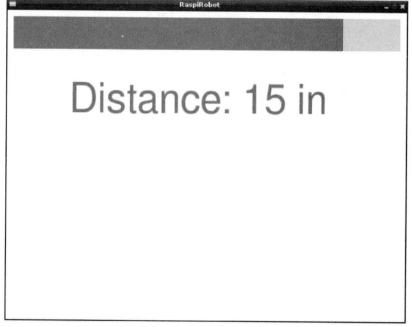

Figure 11-10 *The RaspiRobot display*

The distance is measured and a message is constructed into a surface that is then blitted onto the display. The graphical representation is created by drawing a fixed-size green rectangle and then drawing a red rectangle on top of it whose width depends on the distance sensed by the range finder.

Summary

This project can be treated as the basis for your own robot projects. The RaspiRobotBoard has two extra outputs that could be used to drive a buzzer or control other electronics. Another interesting way of extending the project would be to write software that allows the robot to spin on the spot, and use the range finder to create a sonar-style chart of the room containing the robot. With a Raspberry Pi camera module and a Wi-Fi dongle, all sorts of other possibilities for tele-presence devices arise!

The final chapter of this book looks at where to go next with your Raspberry Pi and provides some useful pointers to other Raspberry Pi resources.

12

What Next

The Raspberry Pi is a phenomenally flexible device that you can use in all sorts of situations—as a desktop computer replacement, a media center, or an embedded computer to be used as a control system.

This chapter provides some pointers for different ways of using your Raspberry Pi and details some resources available to you for programming the Raspberry Pi and making use of it in interesting ways around the home.

Linux Resources

The Raspberry Pi is, of course, one of many computers that runs Linux. You will find useful information in most books on Linux; in particular, look for books that relate to the distribution you are using, which for Raspbian will be Debian.

Aside from the File Manager and applications that require further explanation, you'll want to know more about using the Terminal and configuring Linux. A useful book in this area is *The Linux Command Line: A Complete Introduction,* by William E. Shotts, Jr. Many good resources for learning more about Linux can be found on the Internet, so let your search engine be your friend.

Python Resources

Python is not specific to the Raspberry Pi, and you can find many books and Internet resources devoted to it. For a gentle introduction to Python, you might want to pick up *Python: Visual QuickStart Guide,* by Toby Donaldson. It's similar to this book in style, but provides a different perspective. Also, it's written in a friendly, reassuring manner. If you want something a bit more meaty, but still essentially a beginner's text, consider *Python Programming: An Introduction to Computer Science,* by John Zelle.

When it comes to learning more about Pygame, you'll find *Beginning Game Development with Python and Pygame,* by Will McGugan, to be quite helpful.

Finally, here are some good web resources for Python you'll probably want to add to your browser's favorites list:

- **http://docs.python.org/py3k/**   The official Python site, complete with useful tutorials and reference material.

- **www.pythonware.com/library/tkinter/introduction/**   A useful reference for Tkinter.

- **http://zetcode.com/gui/tkinter/layout/**   This tutorial sheds some much needed light on laying out widgets in Tkinter.

- **www.pygame.org**   The official Pygame site. It contains news, tutorials, reference material, and sample code.

Raspberry Pi Resources

The official website of the Raspberry Pi Foundation is www.raspberrypi.org. This website contains a wealth of useful information, and it's the place to find announcements relating to happenings in the world of Raspberry Pi.

The forums are particularly useful when you are looking for the answer to some knotty problem. You can search the forum for information from others who have already tried to do what you are trying to do, you can post questions, or you can just show off what you've done to the community. When you're looking to update your Raspberry Pi distribution image, this is probably the best place to turn. The downloads page lists the distributions currently in vogue.

The Raspberry Pi even has its own online magazine, wittily named *The MagPi.* This is a free PDF download (www.themagpi.com) and contains a good mixture of features and "how-to" articles that will inspire you to do great things with your Pi.

For more information about the hardware side of using the Raspberry Pi, the following links are useful:

- **http://elinux.org/RPi_VerifiedPeripherals**   A list of peripherals verified as working with the Raspberry Pi.

- **http://elinux.org/RPi_Low-level_peripherals**   A list of peripherals for interfacing with the GPIO connector.

- **www.element14.com/community/docs/DOC-43016/**   A datasheet for the Broadcom chip at the heart of the Raspberry Pi. (This is not for the faint of heart!)

If you are interested in buying hardware add-ons and components for your Raspberry Pi, Adafruit has a whole section devoted to the Raspberry Pi. SparkFun also sells Raspberry Pi add-on boards and modules.

Other Programming Languages

In this book, we have looked exclusively at programming the Raspberry Pi in Python, and with some justification: Python is a popular language that provides a good compromise between ease of use and power. However, Python is by no means the only choice when it comes to programming the Raspberry Pi. The Raspbian Wheezy distribution includes several other languages.

Scratch

Scratch is a visual programming language developed by MIT. It has become popular in education circles as a way of encouraging youngsters to learn programming. Scratch includes its own development environment, like IDLE for Python, but programming is carried out by dragging and dropping programming structures rather than simply typing text.

Figure 12-1 shows a section of one of the sample programs provided with Scratch for the game *Pong*, where a ball is bounced on a paddle.

C

The C programming language is the language used to implement Linux, and the GNU C compiler is included as part of the Raspbian Wheezy distribution.

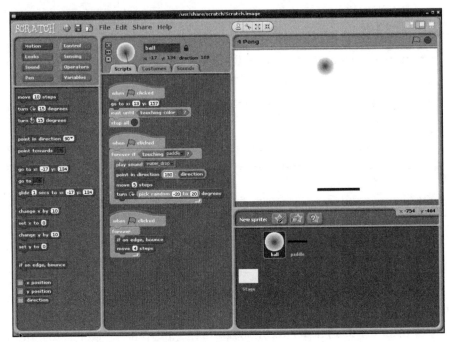

Figure 12-1 *Editing a program in Scratch*

To try out a little "Hello World'" type of program in C, use IDLE to create a file with the following contents:

```c
#include<stdio.h>
main()
{
    printf("\n\nHello World\n\n");
}
```

Save the file, giving it the name hello.c. Then, from the same directory as that file, type the following command in the terminal:

```
gcc hello.c -o hello
```

This will run the C compiler (gcc), converting hello.c into an executable program called just hello. You can run it from the command line by typing the following:

```
./hello
```

The IDLE editor window and command line are shown in Figure 12-2, where you can also see the output produced. Notice that the \n characters create blank lines around the message.

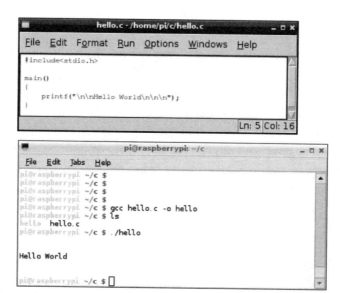

Figure 12-2   *Compiling a C program*

Applications and Projects

Any new piece of technology such as the Raspberry Pi is bound to attract a community of innovative enthusiasts determined to find interesting uses for the Raspberry Pi. At the time of writing, a few interesting projects were in progress, as detailed next.

Media Center (Raspbmc)

Raspbmc is a distribution for the Raspberry Pi that turns it into a media center you can use to play movies and audio stored on USB media attached to the Pi, or you can stream audio and video from other devices such as iPads that are connected to your home network. Raspbmc is based on the successful XBMC project, which started life as a project to use Microsoft Xboxes as media centers. However, it's available on a wide range of platforms.

With the low price of the Raspberry Pi, it seems likely that a lot of them will find their way into little boxes next to the TV—especially now that many TVs have a USB port that can supply the Pi with power.

You can find out more about Raspbmc at www.raspbmc.com/about/, you can learn about the XBMC project at www.xbmc.org. All the software is, of course, open source.

Home Automation

Many small-scale projects are in progress that use the Raspberry Pi for home automation, or *domotics* as it is also known. The ease with which sensors and actuators can be attached to it, either directly or via an Arduino, make the Pi eminently suitable as a control center.

Most approaches have the Raspberry Pi hosting a web server on the local network so that a browser anywhere on the network can be used to control various functions in the home, such as turning lights on and off or controlling the thermostat.

Summary

The Raspberry Pi is a very flexible and low-cost device that will assuredly find many ways of being useful to us. Even as just a simple home computer for web browsing on the TV, it is perfectly adequate (and much cheaper than most other methods). You'll probably find yourself buying more Raspberry Pi units as you start to embed them in projects around your home.

Finally, don't forget to make use of this book's website (www .raspberrypibook.com), where you can find software downloads, ways of contacting the author, as well as errata for the book.

Index